Beneficiation and Management of Technological Waste

This book focuses on the present and future trends, technologies, as well as regulations for the reduction, reuse, and recycling of technological waste globally from the perspective of environmental, social, and economic elements of the different connected systems. It also analyses specific technological waste producing materials and product streams in depth and investigates the future of technological waste and its management considering better end-of-lifecycle designs, legislation, and sustainability perspectives.

Key Features:

- Provides comprehensive review of beneficiation and management of technological waste.
- Discusses sustainability in waste management and beneficiation.
- Covers current practices, policies, and solutions governing the existing supply of discarded or waste technological products and by-products.
- Reviews material-centric waste disposal and beneficiation.
- Includes case studies from industry on technological waste generation and handling.

This book is aimed at graduate students and researchers in materials science and engineering, extractive metallurgy, and sustainable and environmental engineering.

Beneficiation and Management of Technological Waste

Opportunities and Challenges

Anurag Bajpai, Arunabh Meshram, and
Krishanu Biswas

CRC Press
Taylor & Francis Group
Boca Raton London New York

CRC Press is an imprint of the
Taylor & Francis Group, an **informa** business

Designed cover image: Anurag Bajpai

First edition published 2025
by CRC Press
2385 NW Executive Center Drive, Suite 320, Boca Raton FL 33431

and by CRC Press
4 Park Square, Milton Park, Abingdon, Oxon, OX14 4RN

CRC Press is an imprint of Taylor & Francis Group, LLC

ISBN: 9781032555386 (hbk)
ISBN: 9781041025245 (pbk)
ISBN: 9781003619680 (ebk)

DOI: 10.1201/9781003619680

Typeset in Times
by Newgen Publishing UK

Contents

Preface

Fast growth of global population and an increasingly interconnected global economy have been pushing demand for high-tech goods and products significantly enough for the last century. This demand has led to rapid technological breakthroughs and advancements on materials processing technologies and product development, primarily due to the third and fourth industrial revolutions. Although this has eased the lives of humans, there is more waste being generated from these goods and products than ever being thought about it. The traditional concept of "take, use and generate waste", typical of the "linear" economic growth model, creates more waste than we use. Subsequently, waste management faces new challenges and problems. However, there has not been enough discussion on this waste stream in a holistic manner, mainly due to the lack of cumulative categorization of different waste streams under this umbrella. We label them as "technological waste" as they predominantly originate from the rapidly changing demographics of technology and are not predominantly limited to a particular type of waste. Despite the fact that several titles and reviews have presented overviews for various technological waste streams and remediation, there is a dearth of a thorough discussion on the technology of waste and its management. This precludes the reader from gaining insights into the technological waste stream as a whole, focusing on only one or two of its sub-streams.

Beneficiation and Management of Technological Waste focuses on a collective representation of technological waste and its sub-streams. It provides the present and future trends, technologies, as well as regulations for the recovery, reuse, and recycling of technological waste globally. It analyses the viewpoints of worldwide technological waste management systems from the perspectives of science, environmental, social, and economic elements of the different connected systems. It also analyses specific technological waste producing materials and product sub-streams in depth and investigates the future of technological waste management in light of better "end-of-life cycle" designs, sustainability, and legislation perspectives. This book also focuses on the significant shift in technological waste management from the cheapest method to dispose of technological waste in an environmentally benign manner to how technological waste may minimize the need for new materials, leading to a sustainable economy. It aims to promote *zero-waste* as a future technological waste management strategy. The discussion of the range of these relevant topics has focused on the following broad themes:

- A comprehensive perspective on the current state of technological waste.
- Existing practices, solutions, policies, and laws governing the existing supply of discarded or wasted technological products and by-products.
- Contemporary cutting-edge technologies for sustainable reuse and management of technological waste.
- Reuse, recycle, and refurbish – science and technological aspects.
- Future perspectives and potential solutions for environmentally responsible treatment of technology waste.

The book has been structured in eight chapters. Chapter 1 provides the global perspective of waste and waste stream as well as environmental impact. Chapter 2 discusses different waste streams and their nature. Chapter 3 elaborates on the technology of waste management. Different types of waste and their management have been discussed in detail in chapters 4–6. Chapter 7 describes sustainable emerging waste management technologies and the book is summarized in Chapter 8. The disposal and management of technological waste is indeed a pressing problem. The situation is ripe for global catastrophe. As a result, the focus of academia and industry is turning toward finding novel approaches of minimizing technological waste and recycling of the waste to bring the materials to mainstream. Products are now being designed and developed with sustainability at the core. The recycling of the existing waste using environment friendly approaches are doing the round in the core technology development of the major multinational companies. Technologies being developed in the labs have been scaled up and put into actual use. Therefore, this book can be useful for gaining a comprehensive understanding of the concept of technological waste and the challenges and opportunities inherent in its effective management. It is a one-of-a-kind resource for both academicians and businesses, and it will enable experts in the field of sustainable waste management to consider new avenues for the circular economy around technological waste management. This book is a good resource for anybody interested in technological waste and sustainable waste management. At the end, the authors will feel happy and elated, if the students, academicians, and industry professionals find this book useful.

The authors would like to dedicate this book to their respected teachers. The teachers have laid the foundation for research for the authors, with their guidance being utmost importance in developing the research culture. AM would like to dedicate this book to Prof. Kamalesh Kumar Singh, IIT (BHU). AB would like to dedicate this book to all his teachers who have ignited the flames of curiosity within me, fostering the love for learning. AB would also like to dedicate this book to his wife, Dr. Khushubo Devi, a constant source of endless inspiration and encouragement. KB would like to dedicate this book to all the teachers who shaped his life by inculcating an inquisitive mindset, ability to take up challenges and most importantly, ability to perform in adverse situation. KB would like to dedicate this book to his family, who stood by him during difficult times.

Anurag Bajpai *MPIE Dusseldorf, Dusseldorf, Germany*
Arunabh Meshram *IIT Kanpur, Kanpur, India*
Krishanu Biswas *IIT Kanpur, Kanpur, India*

Acknowledgments

Writing a book on a subject, which is one of the global challenges, is always challenging. It would not have been possible to successfully publish it without active help from many. We would like to acknowledge their support, co-operation, and active participation in the book preparation, composing chapters and figures.

Krishanu Biswas (KB) acknowledges his research scholars, especially PhD and Masters students, for helping collect the research papers, referencing, and figures. Among all, Mr. Rahul Mitra, Mr. Syed Musthaq, Ms. Shruti Srivastava, Ms Pallavi Roy, in bringing the chapters to reasonably good shape and readable, and Mr. Buddhvir Singh for typing and compiling chapters. KB also acknowledges the Department of Science and Technology, Government of India, for financial support for the research and development activities for last decade. KB would like to thank his family for bringing the concept of zero waste and sustainability during informal discussion.

Arunabh Meshram (AM) acknowledges the support received from the Department of Materials Science and Engineering, IIT Kanpur and Government of India for financial support for conducting research activities at IIT Kanpur. Mr. Ankur Srivastava, Mr. Harish Trivedi, and Mr. Kaushik Roy have been pivotal in providing suggestions and feedback, leading to improvements in the chapters. AM acknowledges their support. Prof. K.K. Singh, IIT (BHU), Varanasi is deeply revered for his constant guidance and mentorship.

Anurag Bajpai (AB) is grateful to the Department of Materials Science and Engineering at the prestigious Indian Institute of Technology Kanpur, as well as the Department of Science and Technology, Government of India, for their invaluable assistance in facilitating the research endeavors at IIT Kanpur. Furthermore, AB wishes to extend deep gratitude to all the members of the Solidification and Nanomaterials lab, Department of Materials Science and Engineering, IIT Kanpur. Their assistance in both the scientific and non-technical capacities was a critical component in producing this book. AB wishes to dedicate this book to his wife, Dr. Khushubo Devi, whose unwavering support and boundless inspiration have been his guiding light throughout this journey.

Anurag Bajpai *Dusseldorf, Germany*
Arunabh Meshram *Kanpur, India*
Krishanu Biswas *Kanpur, India*

About the Authors

Anurag Bajpai is an Alexander-von-Humboldt postdoctoral fellow at Max-Planck-Institut für Eisenforschung GmbH in Düsseldorf, Germany. His expertise lies at the intersection of artificial intelligence and materials science, where he employs cutting-edge techniques to advance the field of material design and development of sustainable advanced alloys. A significant portion of his research also revolves around the sustainable beneficiation and utilization of electronic waste to develop advanced materials such as metallic catalysts, carbon nanomaterials, and recovery of valuable rare earth metals.

Arunabh Meshram is presently serving as Assistant Professor in the Department of Materials Science and Engineering, Indian Institute of Technology, Kanpur. He was awarded his PhD in Metallurgical Engineering from the Indian Institute of Technology (BHU), Varanasi, India, in 2019. His broad area of interest lies in the areas of non-ferrous extractive metallurgy and materials recycling. He has published 19 articles as author and co-author in reputed journals, including six review articles. The publications have H-index and i-10 index of 12 and 14, respectively, with a total of 570 citations.

Krishanu Biswas is currently working as Ranjit Singh Chair Professor at the Department of Materials Science and Engineering, Indian Institute of Technology, Kanpur. Biswas has made a significant contribution towards understanding of the several important scientific and technological aspects of metallurgy and materials engineering. The research work performed by him ranges from development of complex concentrated high-entropy alloys (HEAs), understanding the solidification behavior of complex concentrated alloys, novel processing of ceramic composites, understanding the alloying behavior at nanoscale, development of bulk alloy catalysis for hydrogen energy and environment. His recent research activities involve research on nanocrystalline high entropy alloys (HEAs), sintering of nanomaterials, ionic nanoparticles, graphene etc. His recent work on multicomponent multiprinciple alloys and ceramic is commendable. He has published over 300 papers in international reputed peer-reviewed journals and delivered 75 invited talks at different national and international conferences. He has contributed to international collaborations with Japan, Germany, Taiwan, and the USA. He teaches extensively at IITK on a variety of subjects including: phase equilibria, manufacturing processes, phase transformations, process metallurgy, solidification processing etc. He is a prolific teacher. He has taught a large number of courses at IIT Kanpur and developed courses on NPTEL platform as well as massive open online courses for the benefit of students in the materials science community in the last five years. He has received numerous awards, fellowships, and other forms of professional recognition.

1 Technological Waste and the Modern Waste Stream

Krishanu Biswas

1.1 INTRODUCTION: WHY WASTE?

Waste is commonly defined as unwanted and unusable material. It is something which we think is no longer necessary today for any utilization. Waste is very common in the world today. We generate extremely large amounts of waste today than even 10 years ago. This is because we have been using lots of products, which are meant for "single time" use and hence are thrown in the landfills, water bodies and even railway tracks or roadways after use. This causes extensive environmental damage. According to the United Nations International Panel (UNIRP, 2012) "Rapid urban and industrial growth" in recent decades has placed huge pressure on the world's natural resources, leading to threats of resource scarcity, price unflattening and degraded ecosystem [1].

For almost the last three centuries (1761–2023), the utilization of natural resources has been immense and so extensive that many of non-renewable resources are fast dwindling and hence we need to take a pragmatic viewpoint for further utilization of these resources. It is now clear and evident that when we have had abundant and inexpensive resources, we seem to overlook the impact on the environment. It has widely been observed that the major multinational companies predominantly use the favored "linear model", which allowed them to extract raw materials from the mother Earth and process them to manufacture the required product as well as to sell them to as many customers possible on the global scale [2]. The customers, on a large-scale, use and discard them after the service life. Therefore, for the last 270 years, the economic model follows the straight-line path "take, use and generate waste", leading to generation of massive waste and major health and environmental hazard. This is typical human tendency and the widely used economic model of "produce and sell" led to generation of humongous amount of waste on the landfills, water bodies and even space [3]. This has been causing large- scale environmental damage, affecting ecosystem in the biosphere. Therefore, the so-called "linear model" is no longer sustainable. Added to this is the large population growth, placing more demand on the dwindling natural resources-metals, minerals, ceramics, and the fossil fuels. These non-renewable resources are constantly dwindling, which is another major cause of concern. Therefore, the utilization of waste to recover and reuse the resources has become the main driver for the world today and the concept is "circular economy" is found to be the reality. This also means, we need to seriously think about further use

DOI: 10.1201/9781003619680-1

of linear model and rather adopt "circular economic" model for sustainability. If possible, we need to completely stop using the "linear model" for the business. It is evident that the "linear model" of growth is the major factor in the generation of waste. In the following, we shall discuss why we need to end the "linear model" and adopt the "circular model" of economy for all in order to make use of waste and turning them back to the mainstream as resource.

1.2 LINEAR MODEL OF ECONOMY: PROBLEMS ASSOCIATED WITH IT

Any economy in the world is rated based on the growth of GDP (Gross Domestic Product), the higher is the growth of GDP; the higher is the economic potential. This is also connected to lending rate. Therefore, every economy would like to grow with a decent rate (4–6%) of growth of GDP. However, GDP growth is connected to resource consumption. It has been indicated that the resource utilization increases by 0.5% for every 1% increase in GDP [4]. A simple calculation will provide the demand on the resources available on the Earth. Let us consider a 25-year timeline (1975–2000). GDP has grown about 225%. Figure 1.1 shows the population growth of the world with projection to 2050 and 2100. The population in the world has grown 64% (4.7

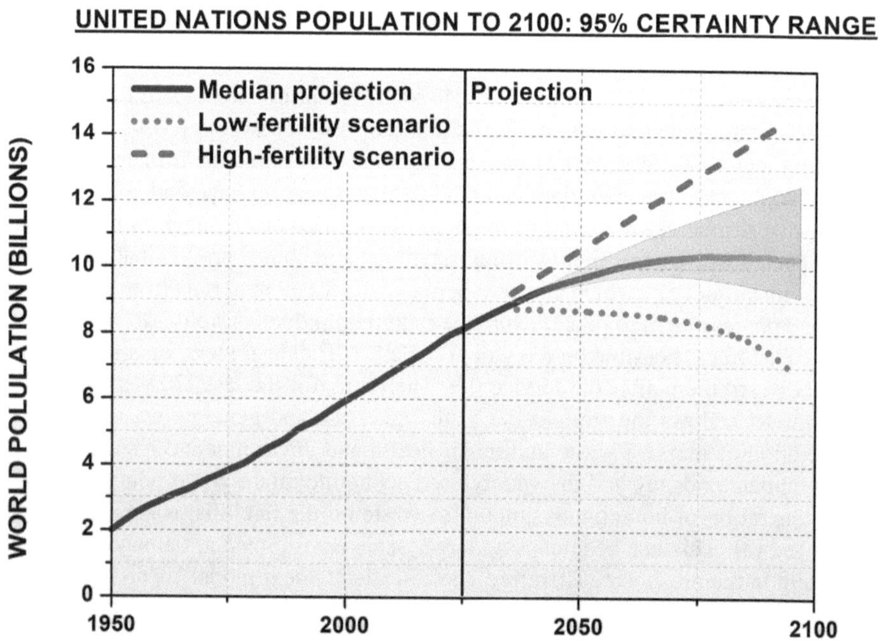

FIGURE 1.1 World population till 2020 and further projection to 2100. The population is expected to reach 11 billion, which will require resources more than 1.5 mother Earth can provide. Consequently, the demand on both renewable as well as non-renewable resources will be extensive and hence, "linear growth" model is no longer sustainable [4].

to 6.97 billion). It is evident the population will grow steadily and reach 16 billion by 2100. This can cause almost 120% or more growth of material consumption. Global population in 2023 is about 8 billion and it is so high the Earth has been struggling to cope up with resources. The environmental damage due to "linear model" has even added to climate change and loss of biodiversity; major long-term effects on our planet. Therefore, it is evident that utilization of net resources is kept on rising with rapid rate. This strained relationship between available resources and utilization has led to rapid increase in the commodity price [4]. Therefore, ever growing population and expanding middle class are expected to drive more demand on resources than the technological innovation and the linear model is no longer tenable. The multinational companies need to alter their strategies to adopt circular economy [5]. Otherwise, the continuation of linear model will definitely lead to multiscale damage of the ecosystem and disrupt the supply chain. Let us now discuss some of the reasons why the linear model is incompatible with the current situation.

The first important aspects are related to the non-renewable resources. Recently, we observed that the demand of various non-renewable resources has been increasing rapidly due to steep increase of population, the data analyses indicate that in last 20 years (2000–2020), the demand for non-renewable sources (fossils fuels, metals, minerals, etc.) has increased almost 70% to be specific, the demand for construction materials being at the highest rate (75%) [4]. This is obviously due to the need for construction materials in the countries belonging to emerging economies, like China, India, Bangladesh, Indonesia, etc. It is beyond doubt that huge rise in demand conjunction with dwindling of reserve of the non-renewable resources is making it extremely difficult to meet the increasing demand. The detailed study also reveals that major non-renewable resources, oil, and metals (Cu, Ag, Au, Co, Pb, Sn) risk running out within the 100 years [6]. This study takes into the consideration of possible discoveries of new reserves. It is indeed observed that, the reserves have been declining, leading to escalation of cost of various resources. Figure 1.2 shows the demand-supply gap for strategic materials and energy. It is evident, the gap widens by 2050 due to dwindling resources of these materials and excessive demand. In a nutshell, non-renewable resources are causing humanity to go to "sleepless night" for next few decades.

In fact, the matter is really worse for some key renewable resources. One of these is drinking water; population explosion in the emerging economies has led to huge increase in global water demand. A detailed study reveals that the demand is stated to increase by 55% between 2000 and 2050 [6]. In this regard a new term, called water stress is utilized. It is termed as a situation in which water resources in a region is insufficient for use. It has also been reported that the ground water, one of the primary sources of safe drinking water has been increasingly polluted. In addition, the rate of depletion of ground water is increasing, placing the 4 billion populations under water stress. The yearly loss of ground water is more than the discharge of major rivers of the water.

Lastly, the key aspect of the transgression of many planetary boundary conditions for sustainable living has been changing. Detailed studies [7] reveal that we have already crossed at least three such boundaries- loss of biodiversity due to high

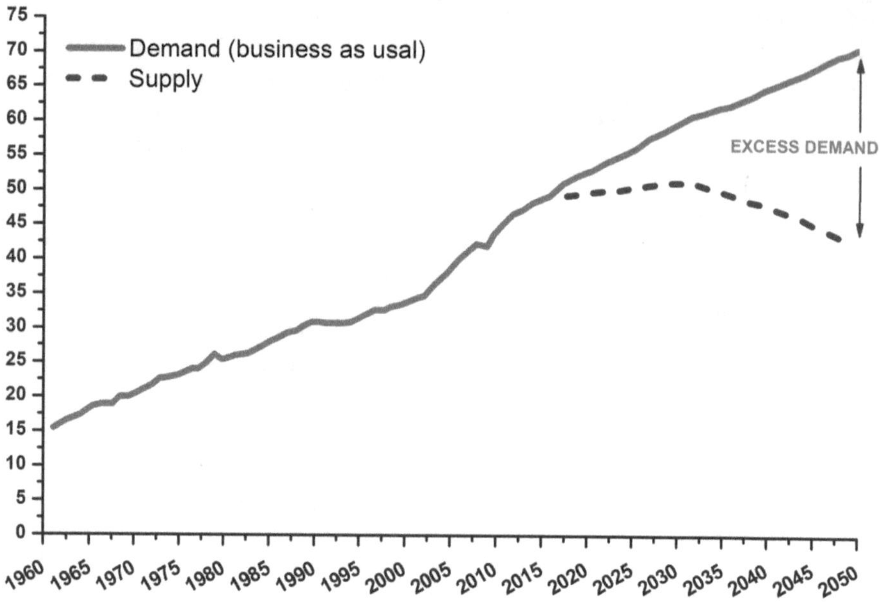

FIGURE 1.2 Supply-Demand gap for strategic materials and energy in the world (1960–2050) [12]. By 2050, the gap widens, indicating resource depletion and excessive demand.

extinction rate, the carbon footprint or atmospheric CO_2 level and atmospheric N_2 fixation. The biodiversity has seen a nosedive; a fall of 50% in the last 50 years (1970–2020) [7]. The CO_2 level in the atmosphere (421 ppm in May 2022) has increased over 50% from the pre-industrial level (280 ppm). This has led to massive changes in our ecosystem, crop production, and even rainfall.

Hence, these aspects have been impacting out "planetary sinks" including forests, oceans, which can absorb these toxic wastes. The emission of CO_2 can not only increase carbon levels in the atmosphere, but also increase the chance of acid rain, threatening the land and sea life. These "sinks" are proved to be critical to cleanse the "dirt" created by us. However, the scarcity of these "sinks" are evidently observed. A recent survey [8] reveals that about 60% of the ecosystem services have been downgraded or even stopped due to unsustainable usage. This led to the rate of extinction of species and loss of biodiversity. We not only need to revive the existing "sinks", but also need to create new "sinks" to take care of the increasing generation of "toxic wastes".

1.3 GENERATION OF WASTE AND IMBALANCE

The earlier discussion has shown that the linear economic model of produce and sell is no longer tenable in the world today. Therefore, we need to apply the principle of 3R, – reduce, reuse, and recycle and rather move to circular economy. It is now proved that the linear economic model generates more trash them we can process. In the world about 11 billion tons of waste are generated annually and only 25% of this waste is reprocessed and returned to the production system. A vast majority (75%) of the waste goes to landfills and oceans. Figure 1.3a shows the trend of generation

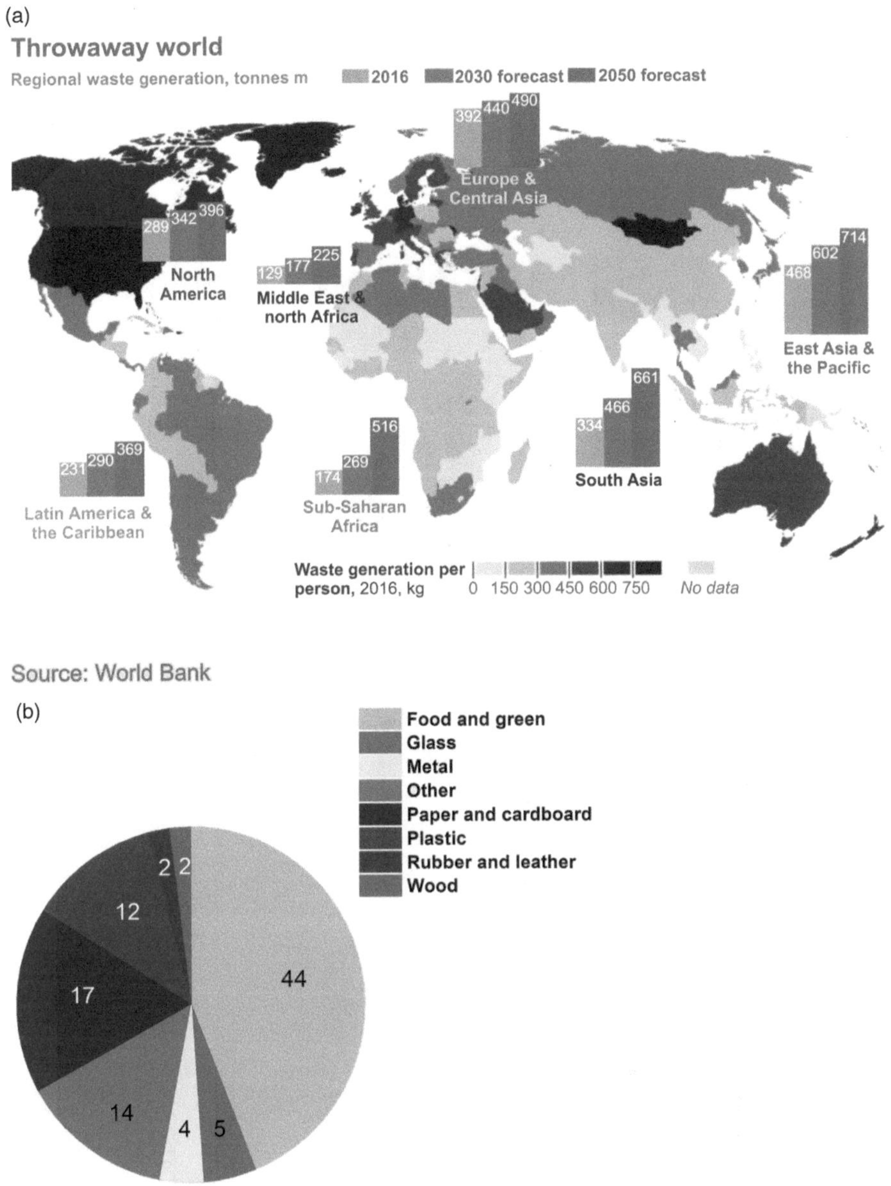

FIGURE 1.3 (a) Waste generation in the world today and further forecast to 2030 and 2050; (b) Composition of waste; food and green have lion's share followed by paper and plastics. *Source*: The Economist and The World Bank.

of wastes in the world as well as projections till 2100. It is evident that the industrial waste in expected to grow by 35% whereas the municipal waste may grow by 75% by 2025 [9]. The trend of growth is highly positive. The global municipal waste is projected to be 3.4 billion tones by 2050 [9]. Figure 1.3b reveals the composition of

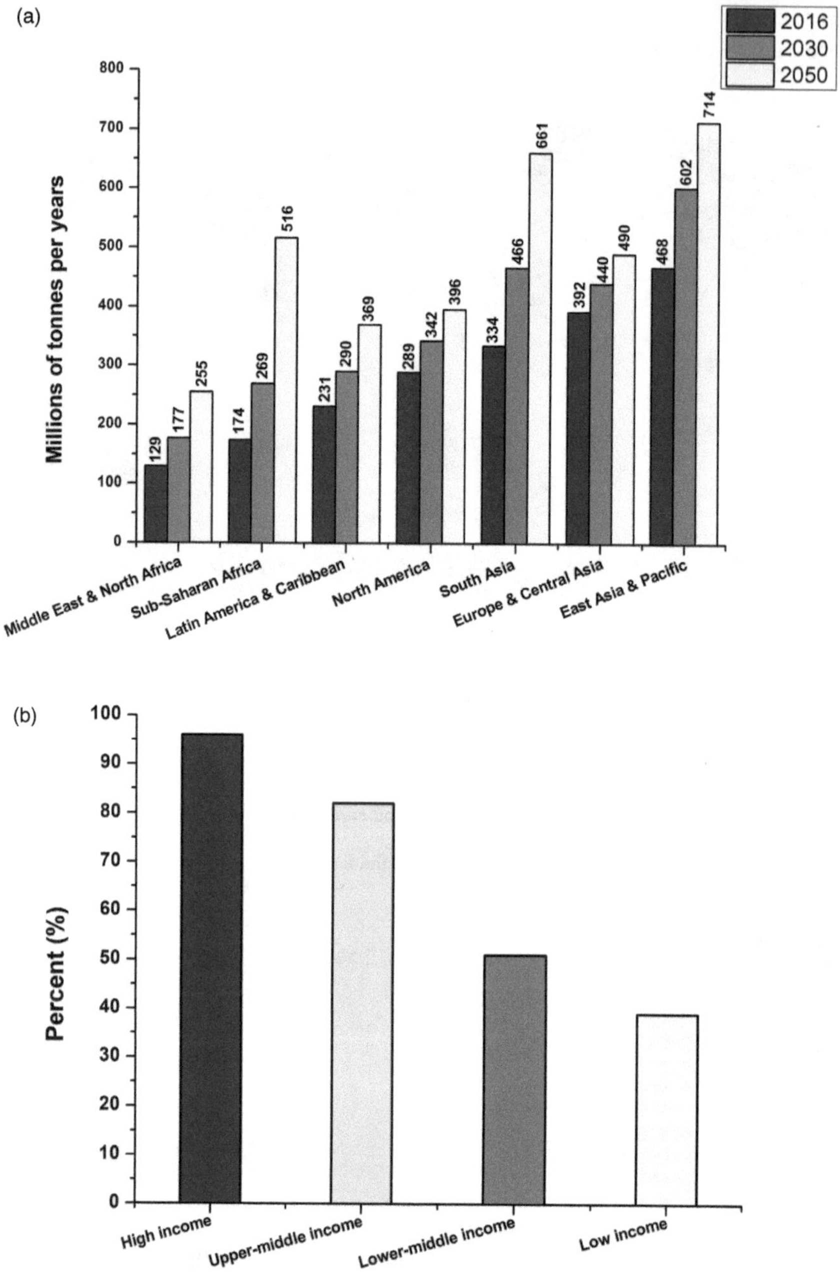

FIGURE 1.4 (a) Statistics on waste generation: waste generation in different regions of our planet. Data in 2016 are on actual and further projections are made for 2030 and 2050. Some regions- Sub-Sharan Africa, South Asia and East Asia and Pacific are expected to see massive rise in waste generation. North America and Europe are expected to show a marginal rise. (b) Average waste collection rate for different income groups. High and Upper-middle income groups generate majority of the waste.
Source: The World Bank.

the waste, food and industrial waste forming the lion's share. The waste generation is not only an environmental problem but also a burning economic issue. The waste in the world today translate to about 1 trillion US$ lost annually. It is equivalent to $300 million of municipal and $700 million of industrial waste. Therefore, the global recycling rate, which is very low today, needs to drastically increases in order to balance the trash generation and recycling. It is clear that steep rise of recycling is the need of the hour to relieve the pressure on the linear growth model. The higher recycling rates lead to less waste going into landfills, water bodies and space. At the same time, higher recycling rate does not preclude the generation of waste, which, at the first needs our attention. Therefore, reduction in the use of resources is possible not only by recycle but also reduce and reuse. The prevention of waste generation by refurbishing as well as reusing the existing products will add much less to ever increasing waste. We also need to understand that we require resources to break down the waste product and build up again to a new product. In a nutshell, waste generation needs to be reduced drastically and whatever is generated needs to be recycled in order to balance the utilization of resources and generation of waste.

1.4 STATUS OF TECHNOLOGICAL WASTE: GLOBAL SCENARIO AND IMPACT ON ENVIRONMENT AND SOCIETY

It has been shown that the business-as-usual with linear economic model is meant to generate lots and lots of waste, creating a dangerous imbalance between the supply of resources and waste generation. The generation of waste, it seems, is heading for a disaster. Figure 1.4a shows the statistics of waste generation in different parts of the world. Asian and African countries are projected to generate more waste than others in future. The waste generation also depends on the income of family. High- and medium-income families in a country contribute more than low and lower medium-income families. Nonetheless, we will generate more waste in future than we are doing now. Since the available resources in the mother Earth are limited, the extensive utilization is taking us to the breaking point if the current practice of "produce and sell" continues. The current growth model, in fact, treats the impact of generation of waste as a matter for "someone else" to deal with. In addition the environmental damage and climate change are becoming more evident as the Earth's ability to absorb and process the waste has been diminishing year by year. In order to understand and resolve this problem, we need to look at the generation of various waste.

In the following, we shall discuss major types of waste produced in the world today. This is popularly known as waste streams. Figure 1.5 shows the various categories of waste in the world. In general, there are five types of waste stream: (a) municipal solid waste; (b) commercial and industrial waste; (c) construction and demolition waste; (d) liquid waste; and (e) hazardous waste. Municipal solid waste constitutes the largest amount of waste produced in the world today. It consists of hazardous waste as well as waste from residential communities. This kind of waste is diverse – it ranges from glass, newspapers to even electronic items. The majority of this waste is sent to waste management plants. The commercial and industrial waste originate from business houses and industries. This constitutes all forms of

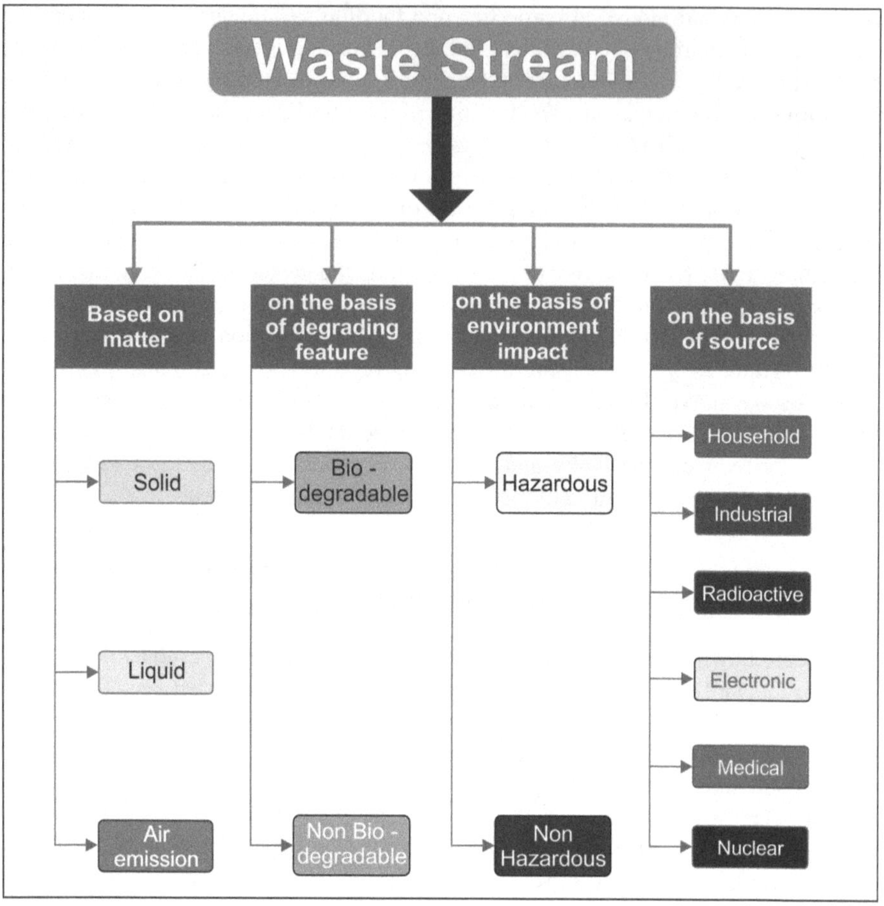

FIGURE 1.5 Different types of waste streams.

business-retail stores to restaurants to even offices. These are many stream paths due to the presence of various starting points. For example, a paper in school can be collected, transported, and sorted into a particular category. Then it can be shredded and pulped; deinked and transformed into new products. On the other hand, the waste from a restaurant follows a different path. The food, which is scrapped off a plate is placed into a trash, going to landfill. Here, it is decomposed and manure is prepared. The industrial solid waste even varies, not between countries at different stages of development, but between the developed and developing countries. In developed countries, the ratio of municipal to industrial waste in 1 to 8. In developing countries, the (Bangladesh, Srilanka, Pakistan, and India) this ratio is 1:3[10]. Nonetheless, the generation of industrial waste has been increasing and it is estimated to have doubled in last 20 years. It is also needed to mention the quality and equality of basic infrastructure in any location in the world will also dictate the waste generation. This is true for both municipal as well as industrial waste. Another important category of waste in construction and demolition waste, which are generated during the process of construction. This includes building of roads, bridges, fly over, subway, or any other construction activity. These wastes are primarily inert and non-biodegradable materials – coverage, sand, metal, plastic, or wood. They are relatively heavy, bulky, and hence occupy a large space. It is common to see piles of these waste lying on the roadside near the large construction projects, causing traffic congestion. It is estimated that approximately 3 billion tons of construction and demolition waste is generated in the world every year. The largest contributions are the emerging economies: China, India, Bangladesh, Philippines, Srilanka, South Africa and collectively they produce about 2 billion tons of the waste. It is evident the construction and demolition waste produce large range of materials-including concrete, glass, wood, metal, and segregation of all the waste materials is difficult due to large variation in composition and complexity of the material. However, wood, glass and metals can be recycled or reused in a few circumstances. The waste management of concrete is extremely difficulty, requiring a unique strategy to recycle. The fourth important category of waste is liquid waste-waste that is formed liquid. These wastes consist of chemical compounds, petroleum products, chemical process waste, medical waste, sewage, septic water, etc. These liquids are hazardous or potentially harmful to human health and environment. Hence, these wastes need to be treated in order to get rid of the harmful things. In big cities, sewage has led to deteriorating water quality of natural water. This has forced the authorities in big cities to set up

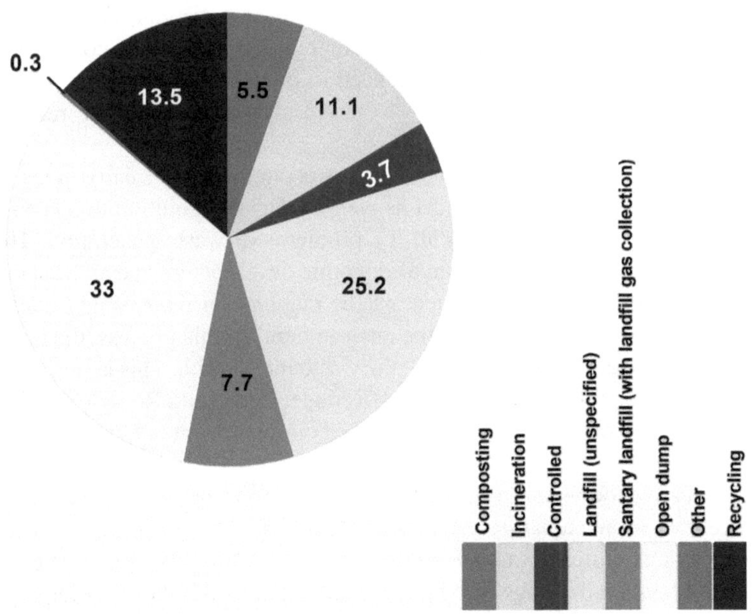

FIGURE 1.6 Statistics on global waste treatment: composting of food and green constitute the highest treatment. Landfill (unspecified, sanitary, and controlled) share almost 50%. *Source*: The World Bank.

wastewater treatment plants to remove the effluent. In fact, the complete removal of waste both from industrial effluent and sewage is possible with restoration of equality water compared to wastewater. The other effluents can also be treated and complete removal of waste is possible. The last category of waste is termed as hazardous waste. These are wastes generated from clinics, hospitals, biological labs, nuclear power plants and these wastes require massive levels of pretreatment before disposal. They potentially cause more hazard than other waste. Waste, depending on type, needs to be treated. However, the majority of the waste generated globally remains untreated and allow to go to landfills, water bodies, or as litter on the surface. However, some waste undergoes treatment and processing in order to add value to it. Figure 1.6 shows the global statistics on treatment of waste. Food waste is predominantly treated well. Almost 50% of the waste generated are kept open without any treatment.

1.5 MATERIALS RECYCLING: ECONOMIC INSIGHT, THE MOUNTING TREASURE

Off-late the transition from linear to circular economy is considered the biggest revolution. It also provides massive economic opportunity. The circular economy will require a change of "supply logic". Obviously, this will need more renewable energies (solar, wind, geothermal etc), a larger number of biomaterials as well as biochemicals which can decompose easily and safely. However, the use of technical materials (metals, ceramics, plastics) must be designed for better recovery and recycling. The design of the products must be done in such a way that we can recover secondary materials as well as recycle the necessary ingredients using low-cost ecofriendly recycling technology. The products even need to be designed for refurbishment and reuse. Hence, the critical supply-side of the circular economy is the reduction of waste and utilization of waste.

The waste, to be specific, adds to the largest economic opportunity in the current situation. Therefore, any technological as well as business solution to convert waste into wealth can provide answers to all the problems of waste generation. This will enable us to grow business and continue economic development and at the same time, it will allow us to do so without affecting the demand on increasingly constrained natural resources and degradation of the environment. In this process, the concept of waste can be eliminated if we convert all the waste into wealth. This underlies the fact that we need to adopt the principle of 3Rs (Reduce, Reuse and Recycle) [11].

Some of the key benefits of adopting 3R's of circular economy can be listed as:

(1) Unlocking the treasure in wasted products by transforming materials management in manufacturing. This can be achieved by reducing direct material use by a significant amount and reusing or recycling the waste. The research carried out on this suggest a huge reward (~US $4.5 billion) for turning waste to wealth by 2030.
(2) The enormous utilization of waste (natural resources, industrial, municipal wastes etc) is possible by removing the ill concept of waste and realizing everything has a value in the world. This can lead to reaping extra profit by

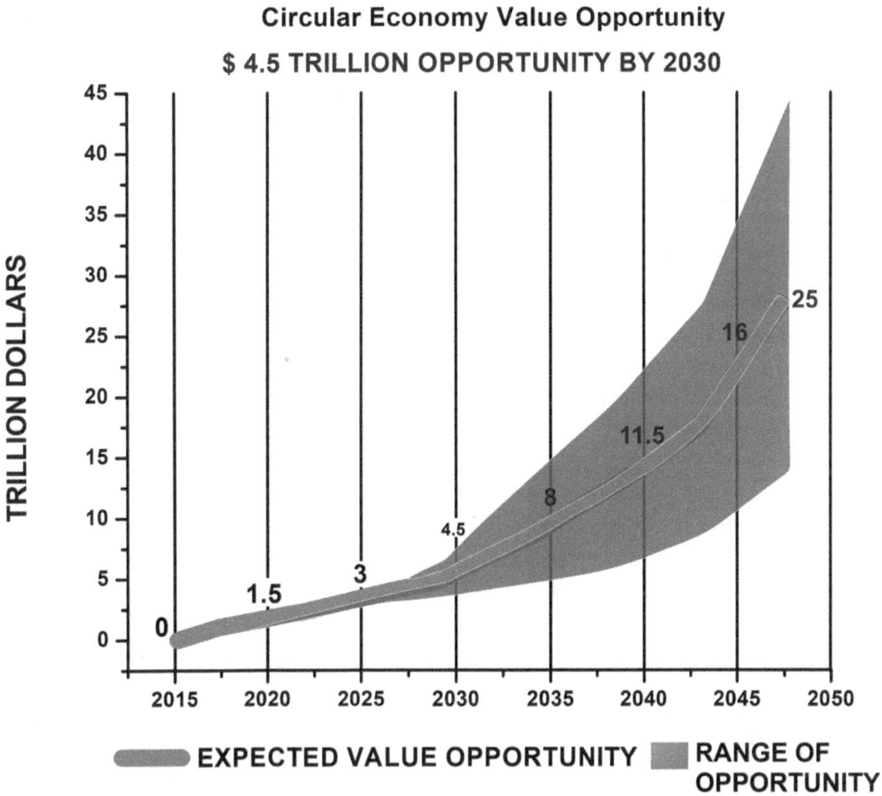

FIGURE 1.7 Economic benefit of adopting circular economy. [12.]

recovering and remanufacturing of the used components and recycling the secondary materials in the waste.

(3) Utilization of the waste will reduce the stress on the dwindling natural resources and create environmentally benign technology, reducing carbon footprint, and may even reverse climate change.

It is evident, all the waste can add up to one of the biggest economic opportunities in the future. Figure 1.7 shows global economic advantage by adopting circular economy. This provides a range of opportunities. By 2050, huge economic value ($4.5 trillion) can be added to economy [12]. Establishing business and technological solutions to transform the waste into wealth enables economic growth without even the need of placing much demand on the increasingly constrained natural resources. This can lead to a performance-based era from the current resource-based era. In this way, the very concept of waste can be eliminated and we can realize that every resource has potential value beyond its current use. The element circular economy – reduce, reuse, and recycle is still not fully understood. However, the concepts of recycling are well known. We understand that many products can be restored, resold, or even refurbished, large number of the products or components of a product need to be recycled to unlock the residual materials and their energy values. Converting waste into wealth or trash into treasure is not new. Across the globe, many countries adopted the conversion of waste food to renewable energy or bio-manure for cultivation. However, such examples are limited to few industries. Major supply chains do not use recycling as a means for circular economy, although this provides lots of advantages.

1.6 CONCLUSIONS AND PERSPECTIVES

Ideally speaking, we need to create zero waste for sustainability of our planet, since there is no planet B. Waste, as mentioned earlier, goes to landfills, water bodies, space, and creates havoc on the nature. The zero waste is economically beneficial also and profitable in the long run. However, the technological viewpoint on zero waste is still a long distance to travel before we can achieve it. It is important that we move in the direction to achieve this goal. The multinational companies need to adopt the circular economy and make sure the gap between supply chain and the waste recycling is reduced. Some of these companies have already adopted the principles of a circular economy in some technological sectors. One of these sectors is renewable energy. However, this effort needs to percolate into all the areas of material use. They can enjoy the "circular advantage" if they spend time to innovate for resource productivity as well as customer value along the full life cycle of a product. However, they need to rethink when it comes to the value delivered to the customer and this will allow them to reconstruct the business model in order to align them to a circular economy.

Technological innovations are going to be major drivers for the circular economy. Since materials form the "lion share" in manufacturing and product development, closing the gap between materials use and recovery from the waste after use will be key in achieving the target of zero waste. This may need smart bin sensors and waste monitoring platforms to optimize route scheduling and the overall logistics of waste collections. This will also need AI-based sorting of the waste for better management. Furthermore, the segregated wastes need to be transported to the processing plants for conversion, recovery and recycle without much addition of carbon footprint. The conversion of these waste into useful products or energy with zero litter will require smart technologies to handle materials. Finally, this will lead to development of new industry, called the "Waste Industry", which needs to implement measures toward a more sustainable future.

REFERENCES

1. "United Nations Department of Economic and Social Affairs/Population Division World Urbanization Prospects: The 2014 Revision," https://esa.un.org/unpd/wup/publications/files/wup2014-report.pdf
2. UNEP, "Responsible Resource Management for a Sustainable World: Findings from the International Resource Panel," 2012, www.unep.org/resource-panel/Portals/24102/PDFs/SYNOPSIS%20Final%20compressed.pdf

3. WWF, "Living Planet Report 2014: Species and Spaces, People and Places," 2014, http://wwf.panda.org/about_our_earth/all_publications/living_planet_report/
4. Stephan Lutter, Stefan Giljum and Marko Lieber, "Global Materials from Flow Database", Accenture Analysis, March 2014.
5. European Commission Report, "Moving Towards to Circular Economy", https://environment.ec.europa.eu/strategy/circular-economy-action-plan_en
6. National Geographic Report, https://education.nationalgeographic.org/resource/nonrenewable-resources/
7. Millennium Ecosystem Report, www.millenniumassessment.org/en/index.html
8. UN Report, Millennium Ecosystem, www.unep.org/resources/report/ecosystem-and-human-well-being-framework-assessment
9. "United Nations Economic and Social Commission for Asia and the Pacific Report", www.unescap.org/sites/default/files/CH08.PDF
10. Rockstrom et al., "Planetary Boundaries: Exploring the Safe Operating Space for Humanity", Ecology and Society 14(2): 2009, 32.
11. United State Environment Protection Agency, www.epa.gov/recycle
12. Douglas MacMillan, Sam Schechner and Lisa Fleisher, "Urbn Snas $41 Billion Valuation", The Wall Street Journal, December 5, 2014.

2 Categorisation of Technological Waste

Anurag Bajpai

2.1 INTRODUCTION

Due to an increasing reliance on technology, the modern era has seen an unprecedented increase in technological advancements, resulting in significant environmental and health issues. A by-product of this technological advancement has also opened up a new branch of waste called "Technological waste". It essentially, refers to a wide range of discarded items resulting from our daily interactions with modern gadgets, industries, and infrastructure. The vast expanses of industrial metallurgical waste, chemical waste, nuclear waste, as well as electronic waste (also known as e-waste) present a variety of intricate challenges that require meticulous scrutiny and implementation of appropriate protocols.

The current scenario and future prospects surrounding technological waste and its different categories are alarming, to say the least. The worldwide waste environment is experiencing a spike in the amount of electronic waste, including abandoned electric and electronic equipment such as computers, mobile phones, and household appliances (Misra, Kumar, and Jain 2021). A staggering amount of 59.4 million metric tonnes of electronic waste was generated worldwide during the calendar year of 2022, based on the findings of the prestigious Global E-waste Statistics Partnership (Sharma et al. 2020). The majority of this enormous amount of e-waste was not properly recycled, illustrating a significant imbalance between its generation and proper disposal. Improper disposal procedures not only cause environmental contamination but also pose significant health risks due to the presence of toxic compounds (Rautela et al. 2021). Similarly, the negative impact of metal extraction and metallurgical waste, which is generated as a by-product of various metal extraction and production operations, is a crucial catalyst for increasing environmental contamination. World Bank estimates that diverse businesses generate more than 4 billion metric tonnes of metallurgical waste from different sources each year globally (Kumar, Holuszko, and Espinosa 2017). The by-product (s) of metal extraction-based industrial operations must be meticulously classified into two separate categories: hazardous and non-hazardous entities. This critical identification can ensure proper and responsible garbage disposal, hence limiting any potential environmental repercussions. Furthermore, the need for a specific framework for the categorisation of nuclear waste stems from its persistent radioactivity, which remains over long periods of time. The

DOI: 10.1201/9781003619680-2

prestigious International Atomic Energy Agency promptly revealed in 2021 that a shocking amount of around 450,000 metric tonnes of high-level nuclear waste had gathered on a global scale (Chapman 2006). To maintain the safety and security of these dangerous materials, this massive waste buildup needs the immediate adoption of durable and impermeable long-term storage solutions. Industrial chemical waste, an amalgamation of hazardous compounds, poses significant threats to ecosystems and human health. According to the United Nations Environment Programme, a whopping 400 million metric tonnes of dangerous chemicals are generated globally each year (Brown 2003). This needs the use of specialised procedures and treatment approaches to mitigate its negative effects on both ecosystems and human well-being.

The classification or categorisation of technological waste using the outline discussed above can serve various functions. It can provide a systematic framework for understanding the composition, dangers, and differentiating characteristics of various waste streams. Implementing a thorough categorisation system allows policymakers and waste management authorities to successfully create and execute policies and frameworks specially targeted to promote safe and sustainable waste management practices (Croff, Richmond, and Williams 1995). Furthermore, waste categorisation improves the capacity to discover prospective recycling channels and delves into the area of cutting-edge technology targeted at recovering useful resources from wasted materials.

In this chapter, an in-depth analysis of several aspects of technological waste categorisation has been provided. The discussion highlights the critical need for a comprehensive and effective approach to waste stream management. Figure 2.1 provides a detailed outline of different ways in which technological waste can be categorised. This includes classification based on source, composition, economic value, disposal method, and hazard level. However, the present discourse focuses on the categorisation of technological waste based on its source. This classification can propel an objective assessment of the issues and opportunities surrounding its management from the scientific perspective. We can effectively propel ourselves towards the realisation of a society that is not only environmentally conscious but also adept at optimising its finite resources through meticulous examination and scrutiny of these diverse technological waste streams, coupled with a comprehensive exploration of the inherent perils they pose and the subsequent implementation of sustainable methodologies.

2.2 METAL EXTRACTION AND BENEFICIATION WASTE

A number of physical and chemical processes are involved in the metal extraction and beneficiation umbrella (Whitworth et al. 2022). These processes and techniques are used to extract and recover valuable metals from naturally occurring mineral deposits. Ore beneficiation refers to a set of intricately defined operations aimed at improving the inherent quality of the ore and increasing its intrinsic value by removing impurities and amplifying its metal concentrations (Ruan, He, and Chi 2019; Houot 1983; Ding, Liu, and Pang 2013). Metal extraction and ore beneficiation are tightly linked processes that rely largely on fundamental concepts of mineral processing and metallurgy. A variety of criteria, including ore composition, mineralogy, and the target metals' characteristics, influence the selection of certain procedures and process parameters. However, they usually include techniques like froth flotation, magnetic

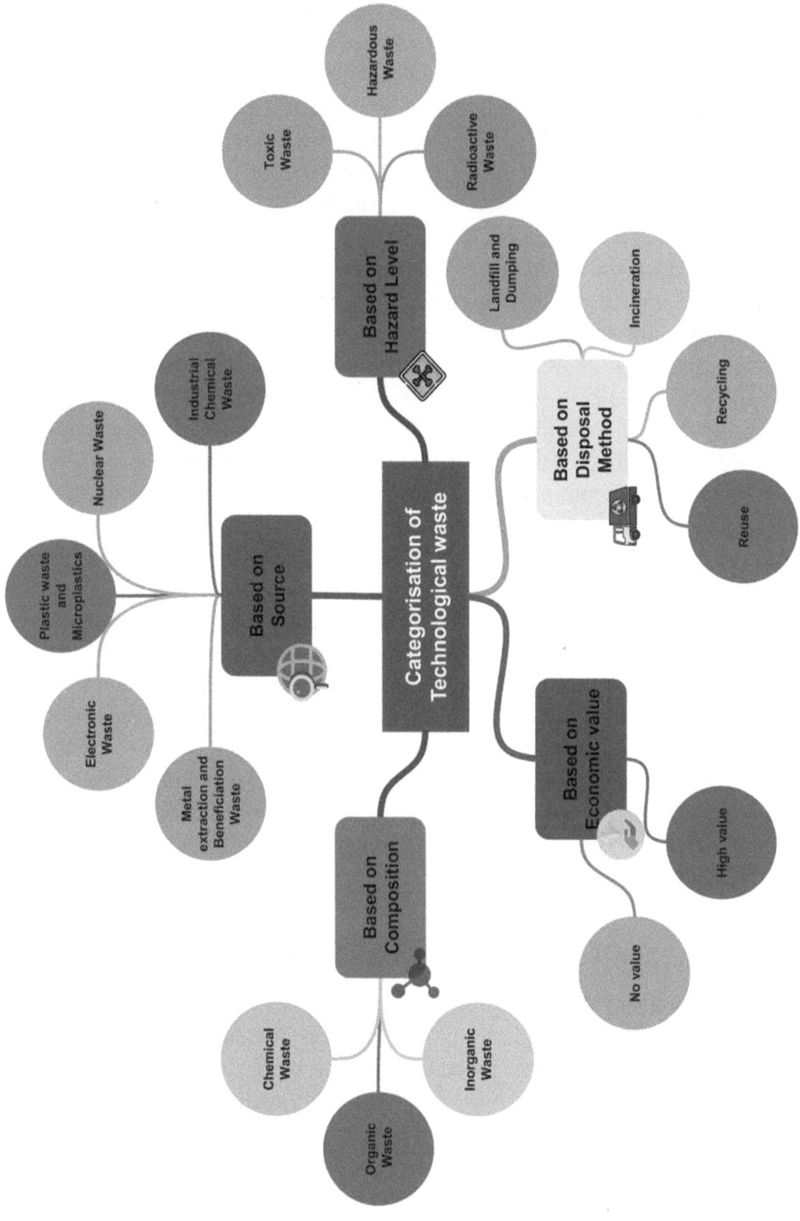

FIGURE 2.1 Categorisation of technological waste into different subcategories based on different basis.

TABLE 2.1
Metal extraction techniques/processes suitable for extracting the selected critical metals from solid mining

Critical Metal	Metal Extraction Processes
Cobalt	Atmospheric leaching, bioleaching, phytomining, pressure leaching, calcine, resin-in-pulp, solvometallurgy
Gallium	Atmospheric leaching, bioleaching
Germanium	Atmospheric leaching, bioleaching
Indium	Bioleaching, solvometallurgy
Lithium	Atmospheric leaching, bioleaching, pressure leaching
Nickel	Atmospheric leaching, bioleaching, phytomining, pressure leaching, resin-in-pulp, solvometallurgy
Rhenium	Atmospheric leaching, bioleaching, pressure leaching
Tellurium	Bioleaching, pressure leaching, solvometallurgy
Tin	Pressure leaching, roasting, smelting, solvometallurgy
Tungsten	Atmospheric leaching, bioleaching, pressure leaching, roasting, smelting
Vanadium	Bioleaching, pressure leaching, solvometallurgy

Source: Whitworth et al. 2022.

separation, and leaching (Das et al. 2007). Table 2.1 summarises some of the metal extraction procedures, or specific processes within them, appropriate for extracting several critical metals from solid mining.

2.2.1 Impact on Ecosystem and Human Health

The procedures associated with metal extraction and beneficiation generate significant amounts of by-products, frequently containing potentially harmful elements such as heavy metals and sulphur compounds. These compounds have the power to permeate the surrounding environment, influencing the quality of the soil, water, and air. Furthermore, the use of chemical compounds in the context of beneficiation techniques, such as reagents and flocculants, has the potential to have negative environmental consequences (Ledin and Pedersen 1996). The lead smelting industry, a major metallurgical processing sector, has long been connected with considerable environmental contamination and negative health effects. Several case studies have shown heightened lead concentrations in the soil, which poses significant health risks to the indigenous population, particularly the vulnerable demographics of children (Laidlaw et al. 2017). Furthermore, recent empirical information, as published by the respected International Resource Panel, has shone a light on a troubling fact about mining and metallurgical beneficiation by-products. Surprisingly, the annual volume of this waste stream is expected to be greater than 30 billion metric tonnes. It is worth noting that the enormous volume of metal extraction waste is tightly tied to the amount of ore produced, with the waste-to-ore ratio demonstrating a wide range, ranging between 1:1 and 5:1. It is critical to emphasise that this ratio is dependent on the specific material being extracted (Oberle et al. 2019). At the same time, the Minamata illness outbreak in Japan provides a sobering reminder of the consequences of poor metal extraction waste management procedures. The introduction of mercury-laden industrial effluents into the aquatic ecosystem of Minamata Bay is another case study, which caused a severe state of mercury contamination, resulting in a cascade of negative consequences, including the manifestation of neurological disorders and the premature death of people living nearby (Eto, Marumoto, and Takeya 2010).

2.2.2 Categorisation of Metal Extraction and Beneficiation Waste

The categorisation and classification of waste products resulting from metal extraction and ore beneficiation operations are critical in improving our understanding of their intrinsic characteristics, possible risks, and appropriate procedures for their effective management (Jamieson, Walker, and Parsons 2015; Salgado-Almeida et al. 2022). The waste generated by the mining and metal beneficiation industry can be majorly divided into three categories: *overburden*, *mine tailings* and *metallurgical slag*.

The development of *overburden* as a result of metal extraction procedures is a significant environmental issue that requires quick and urgent attention. The extraction of metals from ores entails the removal of large amounts of superjacent rock, soil, and other materials, known collectively as overburden (Figure 2.2). The disposal of this residual debris, which typically consists of minerals of poor quality or lacking intrinsic value,

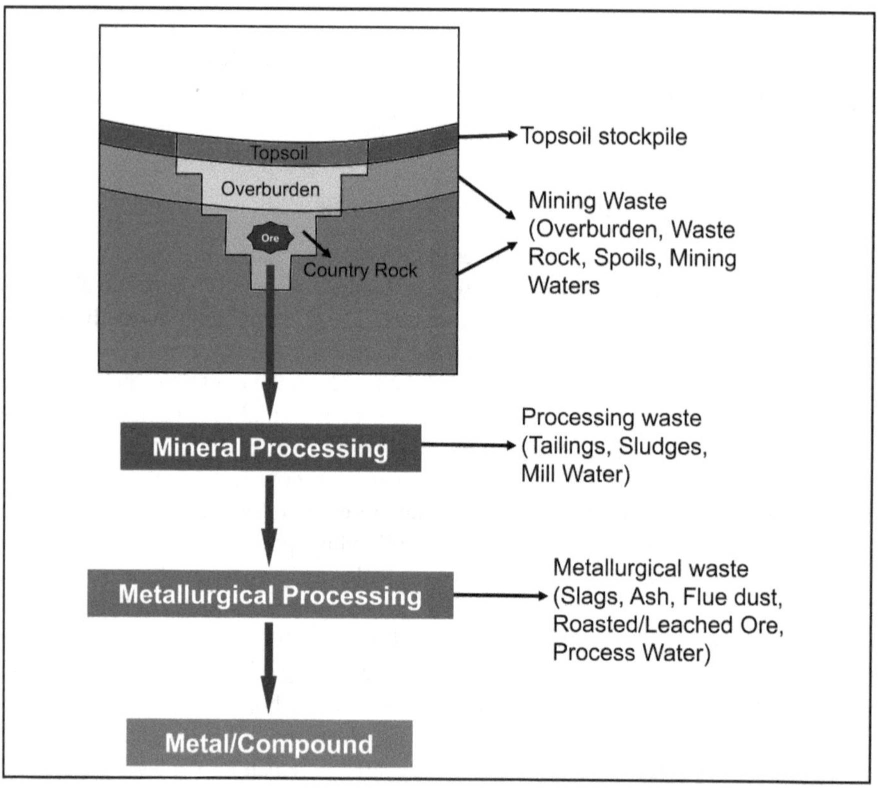

FIGURE 2.2 Schematic illustration of the types of mine waste generated from a metal commodity. (Lottermoser 2010.)

offers various challenges due to its large volume, chemical composition, and proclivity to cause ecological contamination (Lottermoser 2010). The inclusion of potentially hazardous components in overburden, such as heavy metals or sulphides, can have a negative impact on both water quality and ecosystems unless adequate management measures are employed. In recent years, the scientific community has made tremendous progress in understanding the negative repercussions of excessive waste accumulation (Jain and Das 2017). The presence of chemical interactions between mining overburden debris and other components found in the atmosphere can result in acid mine drainage (AMD). This form of drainage is distinguished by high amounts of heavy metals and acidic chemicals. The harmful impacts of AMD go beyond mere contamination of nearby bodies of water, posing major risks to both aquatic creatures and human health (Akcil and Koldas 2006). Furthermore, the physical shift of overburden waste has major ramifications for the landscape, particularly in terms of habitat damage and eventual biodiversity loss.

Mine tailings, or the remaining materials after valuable metals are extracted from ores, are often placed in designated sites such as tailings ponds or dams. Tailings are generally combinations of crushed rock and processing fluids from mills, washeries, or concentrators that remain after metals, minerals, mineral fuels, or coal have been extracted from the mining resource (Figure 2.3) (Lottermoser 2010). These

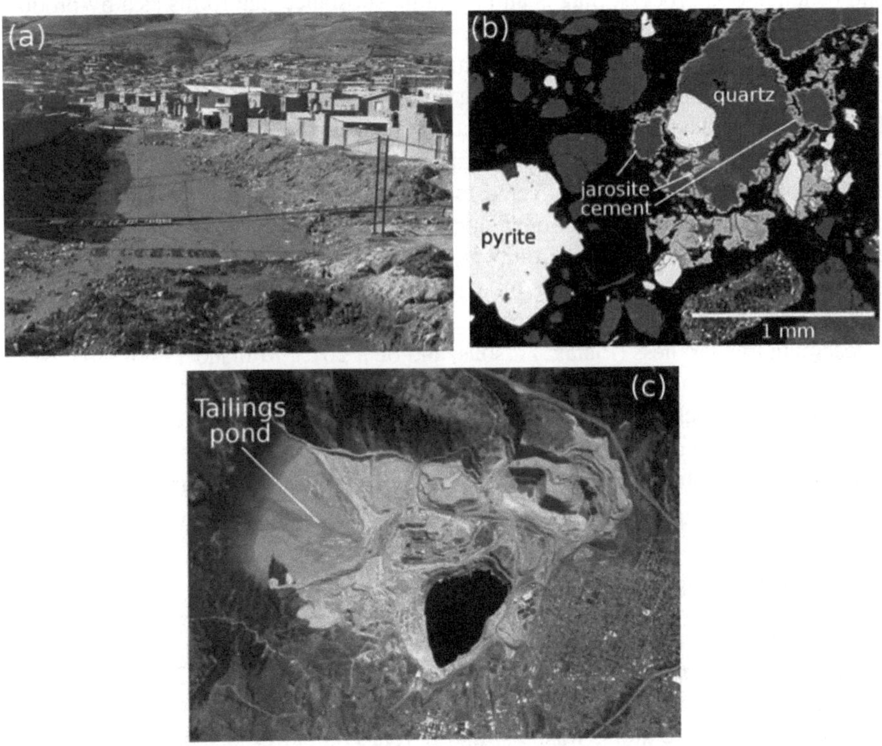

FIGURE 2.3 Photo and photomicrograph of tailings. (a) Grey sulphide tailings in a stream flowing through the town of Huanuni, Bolivia; (b) Bolivian mine tailings primary phases undergoing cementation by secondary jarosite following three years of column leaching experiments; (c) mining area at Berkeley, Montana, USA. Grey tailings are stored in the tailings pond shown to the left of the dark-coloured Berkeley Pit. (Kossoff et al. 2014.)

repositories act as containment hubs for the post-extraction materials. These residues have been discovered to contain significant amounts of metallic elements, including but not limited to copper, lead, and zinc. They may also include various other minerals and chemicals that were used in extraction and beneficiation processes (Jain and Das 2017). Improper or insufficient tailings management techniques can unintentionally release hazardous compounds into nearby ecosystems, ultimately contaminating soil and water resources. Throughout history, tailings dam breaches have been responsible for catastrophic environmental disasters of enormous magnitude (Shengo 2021). Because of the discharge of mine tailings into the local river system, the Ok Tedi mine in Papua New Guinea has been a source of substantial environmental concern. The discharge of mine tailings containing high quantities of copper and other metallic elements has caused significant environmental degradation and harmed the sustenance of indigenous tribes living nearby (Storey and Yarrao 2008). Such occurrences serve as a stark reminder of the critical need to improve and refine waste management techniques.

The separation and concentration of valuable metals from gangue minerals are required for the extraction of valuable metals from raw ores *via* ore beneficiation. *Metallurgical slag* production is a key side step in practically every metal extraction and beneficiation process. Slag is an intriguing substance that forms as a by-product of the complex smelting and refining methods used in the extraction of valuable metals from their respective ores. Non-metallic minerals, contaminants, and fluxing agents make up the majority of the slag composition (Potysz, van Hullebusch, and Kierczak 2018). The global output of metallurgical slag has increased to an estimated 500 million metric tonnes on an annual basis (Kim and Azimi 2021). This large amount of waste has been discovered to have negative consequences on the environment and human health. Improper slag disposal may result in the leaching of hazardous compounds into the environment, thereby impacting soil composition, water quality, and the delicate balance of neighbouring ecosystems. However, the steel industry has made significant advances in recycling and repurposing slag, shining as a beacon in the darkness (Piatak, Parsons, and Seal 2015; Min and Tsukihashi 2017). Tata Steel, a leading steel company in India, has skillfully implemented numerous innovative approaches to successfully harness the residual by-products generated during the steel manufacturing process. The company's use of slag in the production of construction materials helps to reduce the demand for natural aggregates, resulting in a remarkable reduction in the company's environmental footprint.

2.3 ELECTRONIC WASTE

The term "electronic waste," sometimes known as "e-waste," refers to a wide variety of obsolete electronic equipment (Kiddee, Naidu, and Wong 2013). This covers a wide range of electronic gadgets, such as desktops, notebooks, smartphones, tablets, TVs, printers, refrigerators, washers, and dryers (Bajpai et al. 2021). E-waste has become increasingly widespread due to the rapid pace of technological development and the insatiable appetite of modern consumers for the latest cutting-edge technologies in electronic devices. According to research conducted by the respected Global

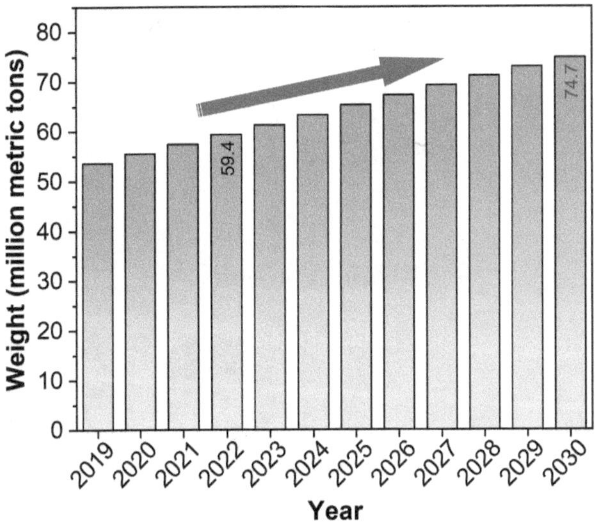

FIGURE 2.4 E-waste generation between 2019 and 2030. (numbers from 2023 to 2030 are projections.) (Shittu, Williams, and Shaw 2021.)

E-waste Statistics Partnership, the worldwide production of e-waste in 2022 was a whopping 59.4 million metric tonnes. This enormous number represents the significant influence of electronic use on our world, with an average of 7.3 kg per person. By 2030, it is estimated that global e-waste production will balloon to an unprecedented 74 million metric tonnes, creating a daunting and urgent problem in the field of sustainable resource management (Sharma et al. 2020). Figure 2.4 summarises the trend of projected increasing amounts of e-waste in million metric tonnes, with the total volume generated expected to rise to 74 million metric tonnes by 2030.

2.3.1 Categorisation of Electronic Waste

Similar to other forms of technological waste, it is paramount to categorise and label different types and forms of e-waste for their effective management. E-waste can be broken down into subcategories primarily depending on the type of discarded electronic item. Precisely, the approach incorporates considerations for the device type, material composition, and recycling possibilities that all contribute to the unique characteristics of e-waste. Everything from computers to mobile phones to televisions and home appliances falls under this categorisation (Perkins et al. 2014). Figure 2.5 illustrates different electric and electronic equipment that fall into different classifications of e-waste based on device type. Besides, the disposal and recycling strategies for each device category must account for its individual traits and components (Terazono et al. 2006). The material composition of e-waste is crucial to its categorisation since it reveals the presence or absence of valuable and potentially harmful elements within the devices. Complex electrical and electronic devices rely on a wide variety of parts and materials that come together to perform their intended

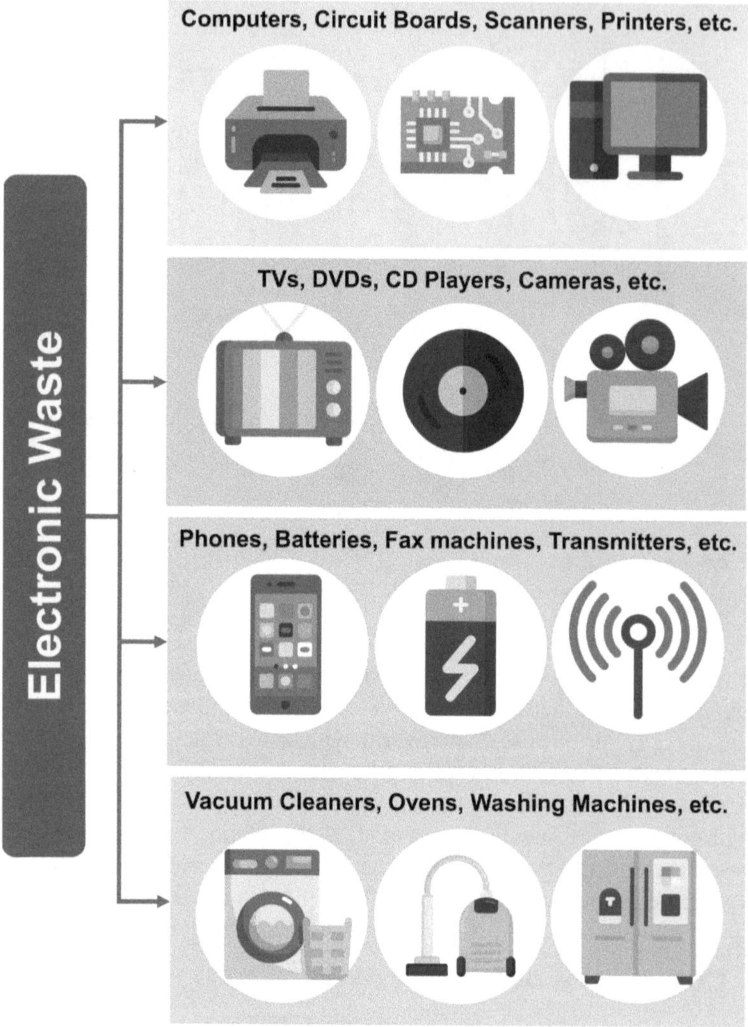

FIGURE 2.5 Different electric and electronic equipment fall into different classifications of e-waste based on device type.

tasks. These elements include metals, polymers, glass, and, unfortunately, toxic compounds that may be harmful to people and ecosystems. In the quest for long-term technological progress, it is important to consider and control potentially harmful substances like lead, mercury, and brominated flame retardants. It is crucial to thoroughly comprehend the complex composition and potential ecological repercussions connected with different types of e-waste to accomplish effective management and recycling methods. In this direction, differentiating potentially harmful electronic trash from non-hazardous garbage is important for handling it properly. The process of collecting and recycling valuable resources like precious metals or rare earth elements can be sped up by sorting e-waste according to its valuable constituents. Gold, silver, and palladium are all precious metals that are well recognised for their significance. Rare earth metals are another class of valuable materials that extend beyond precious metals. These elements have recently attracted much interest due to their rarity and crucial use in many technological applications. They are crucial to the development of cutting-edge devices and systems due to their inherent qualities and one-of-a-kind electrical setups. Accordingly, precious metals and rare earth metals abundantly present in e-waste are prime examples of invaluable substances captivating the imagination and driving scientific exploration in the vast expanse of e-waste management.

2.3.2 Perils on Ecology and Environment

While e-waste is often considered as urban mines rich in valuable metallic and non-metallic constituents, it poses serious risks to human and environmental health due to toxic substances. These effects have worldwide reach and affect countries at all stages of economic growth, from the most developed to the most developing (Awasthi, Zeng, and Li 2016). The insufficient management and disposal techniques applied for toxic compounds inherent to electronic equipment are a major contributor to the disastrous effects that e-waste has on the environment (Beula and Sureshkumar 2021). An estimated 50 million metric tonnes of e-waste are thrown away incorrectly each year, according to the United Nations Environment Programme (UNEP). In congruence, according to data published in the highly regarded publication Global E-waste Monitor 2020, only 17.4 percent of the electronic garbage produced worldwide in 2019 was properly collected and recycled (Forti et al. 2020). This disturbing finding highlights the reality that most of this e-waste is either neglected or managed insufficiently.

The adverse effects of e-waste on the environment and human health have made its proper management and disposal an urgent issue in recent years. The prevalence of this problem may be traced back to the introduction of potentially harmful materials into consumer electronics. Figure 2.6 schematically represents some of the adverse effects of e-waste on humans and ecology. Dangerous compounds cover a wide variety of harmful chemicals, including many different types of heavy metals. Brominated flame retardants and other harmful compounds are also included here. E-waste can contaminate soil, water sources, and air if not handled properly, resulting in health risks for humans and wildlife. The poisoning of ecosystems by e-waste has negative effects, such as altering the soil composition, decreasing water quality, and polluting the air. Heavy metals and hazardous compounds released into the environment pose a serious risk to the stability of plant and animal communities, leading to extinctions and the breakdown of ecosystems (Abdelbasir et al. 2018). Furthermore, the contamination of the complex web of creatures that make up the food chain is a real possibility when dangerous compounds are released into aquatic habitats. When people consume contaminated food or water, it can have negative effects on their health. According to scientific research, the harmful elements in e-waste have been shown to survive in the environment for very long periods of time (Heacock et al.

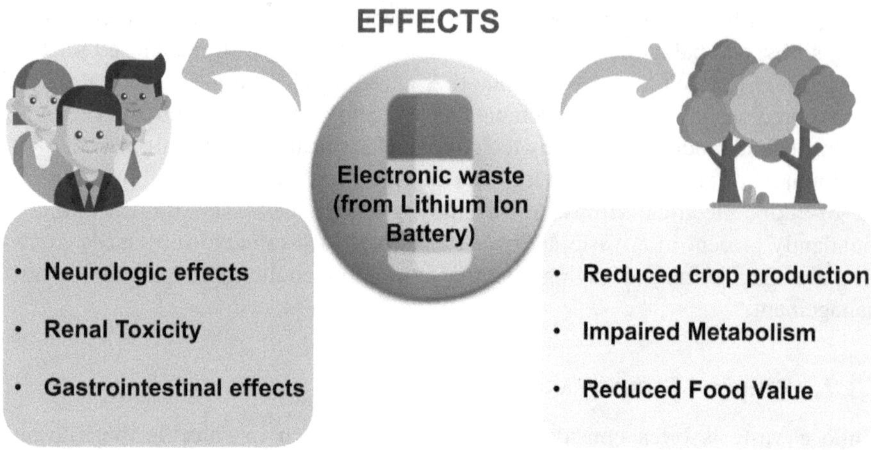

FIGURE 2.6 Schematic describing the adverse effects of e-waste on the environment.

2016). Figure 2.7 presents schematic information on common toxicants released from unsound waste management activities (Yohannessen et al. 2019). Lead is an element that has been shown to be toxic to the central nervous system and hence to the brain development of children and other sensitive populations (Zeng et al. 2020). Bioaccumulation of mercury in aquatic habitats has been examined extensively, demonstrating significant consequences for a wide range of animals. Over time, mercury may build up in living species' tissues, where it can have negative impacts on health and well-being due to its toxicity in higher amounts. The bioaccumulation of mercury also causes alarm because of the threat it poses to human populations through the food chain. The processes behind mercury bioaccumulation in aquatic habitats have been the subject of several scientific studies. It is well-documented that fish, shrimp, and other aquatic species may easily absorb mercury through their gills, skin, or the consumption of polluted food. After being ingested, mercury goes through a complicated series of chemical reactions, one of which is methylation, which turns it into a highly toxic methylmer (Galappaththi and Suraweera 2020; Li and Achal 2020). Brominated flame retardants (BFRs) released during the disposal of e-waste have been linked to the widespread release of POPs, negatively affecting the surrounding ecosystem (Ma et al. 2021).

To make matters worse, inefficient disposal methods like incineration or landfill disposal can add significantly to air and water pollution, hastening the decline of the local environment. Prolonged exposure to e-waste can also have serious consequences for the health of those who work with it and those who live close to it. Informal e-waste recycling businesses, which are most common in underdeveloped countries, expose workers to several health risks due to a lack of proper safety measures (Alabi et al. 2021). Disassembling, recycling, and disposing of e-waste can release toxic airborne chemicals, posing serious health risks to nearby residents. Inadequate safety precautions used during manual disassembly, leading to direct contact with hazardous chemicals, can cause skin diseases, compromise organ function, and cause long-term health problems (Decharat and Kiddee 2020). Several such cases have been reported

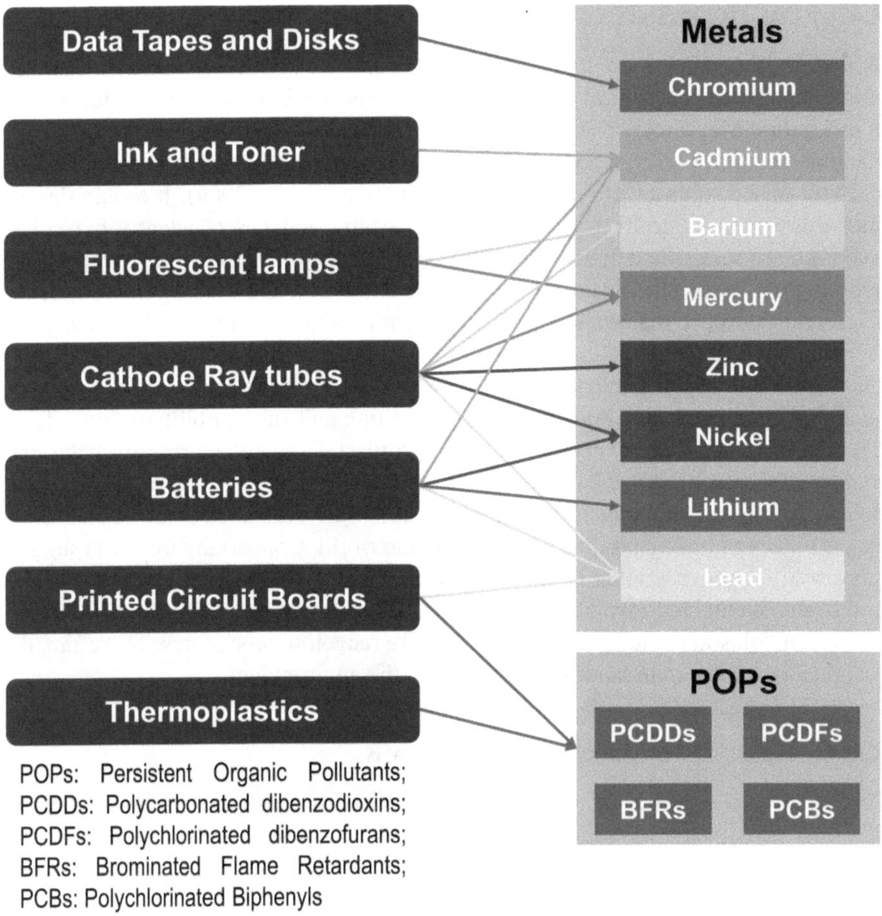

FIGURE 2.7 Common toxicants released from unsound waste management activities. (Yohannessen et al. 2019.)

throughout the globe. Agbogbloshie, Ghana, has been one of the epicentres of e-waste related catastrophes. The aforementioned spot is quickly becoming a world-famous dump for unwanted electronics. Burning cables for copper recovery in the uncontrolled recycling practices at Agbogbloshie has severe ecological and physiological consequences. As a result of prolonged exposure to toxic gases and the presence of large amounts of metals, many people in the workforce and the surrounding community have developed serious health problems (Daum, Stoler, and Grant 2017). In another case, e-waste recycling has reached a critical mass in the Chinese city of Guiyu. Acid leaching and open combustion are two examples of unregulated recycling practices that have contributed to widespread air, land, and water pollution. Multiple field studies have shown without a doubt that there are dangerously high levels of toxic heavy metals in the environment, posing a grave threat to the health of local residents and workers (Huang, Shi, and Wu 2021).

2.4 PLASTIC WASTE

In modern life, plastics are ubiquitous. Its early usage dates back to 1600 B.C., when human hands shaped natural rubber and polymerised it into different useful objects in prehistoric Mesoamerica (Hosler, Burkett, and Tarkanian 1999). Diverse usage and manufacturing of plastics and plastic products began in 1839 when polystyrene (PS) and vulcanised rubber were discovered (Andrady and Neal 2009). It took a decade after this for mass production of plastics to begin, and has constantly expanded. According to scientific definition, plastics are synthetic polymers manufactured from hydrocarbon-based chemicals obtained by the extraction and refining of fossil fuels. These multipurpose materials are carefully synthesised from extended sequences of repeating monomers, which play a major role in determining their unique properties and a wide array of possible uses. The wide variety of plastic formulations results in materials with varying stability, recycling potential, and susceptibility to breakdown (Hees et al. 2019). The exponential growth in plastics manufacturing over the past century has been observed by scientists as a noteworthy occurrence in recent times. Because of this unparalleled growth, the year 2019 saw a record-breaking 368 million metric tonnes of plastic produced throughout the world. Concerning forecasts suggest this already massive amount is set to double by 2050, providing current consumption trends are maintained (Walker and Fequet 2023). Unfortunately, only a small percentage of "plastic waste" gets through effective recycling procedures. Therefore, the vast majority ends up in landfills or as litter in the environment.

2.4.1 Impact on Environment and Humans

The widespread presence of plastic waste in the environment, as well as the damage they do to ecosystems, has become a major cause for concern on a worldwide scale. The effects of plastic waste on the environment are devastating. Figure 2.8 depicts the future course of plastic materials that have an adverse impact on the environment. As a result of human activity, plastic waste is exceptional in its capacity to persist in the environment for very long, hence causing persistent pollution (Ali et al. 2021). An increasingly pressing issue in environmental deterioration is the longevity of plastic waste, which has been around for millennia. Plastics may break through barriers in the environment and contaminate land, water, and air, wreaking havoc on the delicate balance of ecosystems and habitats. A shocking discovery in the field of ecology has thrown light on the plight of the seas. It has been estimated that 8 million metric tonnes of plastic enter the oceans each year and spread over the vast marine ecosystems (Verma et al. 2016). The disturbing increase in plastic waste exacerbates the severe problem of marine contamination. If current trends continue, it is estimated that by the year 2050, there will be more plastic than fish in the waters of Earth. Furthermore, it has been shown that maritime bird species have a predisposition to ingest tiny pieces of synthetic polymers. Malnutrition and reproductive abnormalities are only two of the negative health effects that have been linked to this disturbing phenomenon (Dutta and Choudhury 2018).

To further elaborate on the harmful effects of plastic waste, the release of harmful chemicals into the environment from plastics can occur during production or as a by-product of their natural breakdown. It has been noted that chemicals used in the production of plastic, such as plasticisers and flame retardants, have the ability to leak

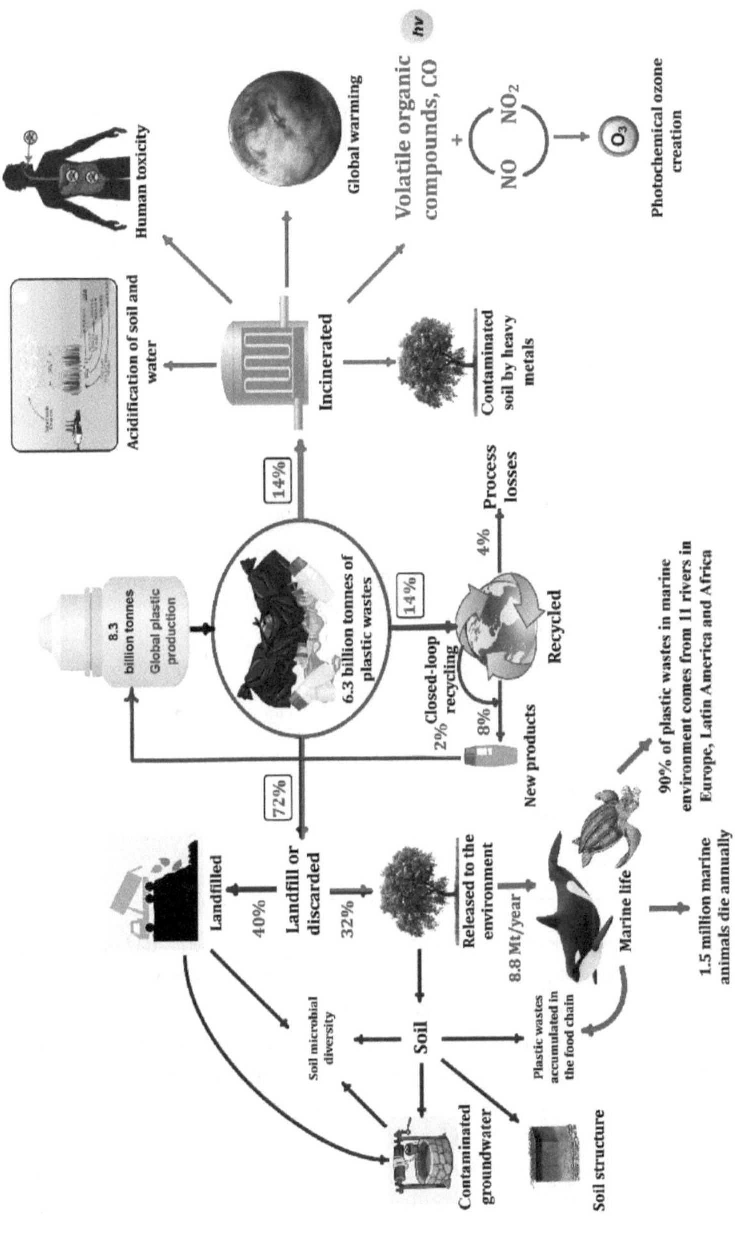

FIGURE 2.8 Illustration showing the fate of plastic materials that negatively influence the environment. (Ali et al. 2021.)

TABLE 2.2
Different additives used in plastic production, their effects and the plastic types

Toxic Additives	Uses	Public Health Effect	Plastic Types
Bisphenol A	Plasticisers, can liner	Mimics oestrogen, Ovarian disorder	Polyvinyl chloride (PVC), Polycarbonate (PC)
Phthalates	Plasticisers, artificial fragrances	Interference with testosterone, sperm motility	Polystyrene (PS), Polyvinyl chloride (PVC)
Persistent Organic Pollutants (POPs)	Pesticides, flame retardants, etc.	Possible neurological and reproductive damage	All plastics
Dioxins	Formed during low temperature combustion of PVC	Carcinogen, interferes with testosterone	All plastics
Polycyclic aromatic hydrocarbon (PAHs)	Use in making pesticides	Developmental and reproductive toxicity	All plastics
Polychlorinated biphenyls (PCBs)	Dielectrics in electrical equipment	Interferes with thyroid hormone	All plastics
Styrene monomer	Breakdown product	Carcinogen, can form DNA adducts	Polystyrene
Nonylphenol	Anti-static, anti-fog, surfactant (in detergents)	Mimics oestrogen	PVC

Source: Halden 2010.

out and contaminate neighbouring ecosystems slowly (Li et al. 2021). In addition to being present everywhere, plastic waste also poses a significant physical threat, becoming entangled in or blocking the complex digestive systems of a wide range of animals. The emission of harmful substances, such as bisphenol A (BPA) and phthalates, has been noted as a major worry in the scientific literature as a possible danger associated with the accumulation of plastic waste. Table 2.2 lists different additives used in plastic production and their effects corresponding to different types of plastics (Halden 2010). The aforementioned compounds may cause endocrine disruption and pose risks to the health of both animals and humans. It is also important to recognise that plastic waste in marine environments can provide a fertile breeding ground for a wide variety of diseases. This might greatly increase the risk of spreading illness among the variety of marine species that call these places home (Thushari and

Senevirathna 2020). The growing buildup of plastic waste threatens the stability of Earth's complex ecosystems and the great diversity of life that exists inside them.

2.4.2 Categorisation of Plastic Waste

Plastics are not naturally biodegradable because they cannot be broken down into their component parts by microorganisms or other organisms. In contrast, when exposed to sunlight, these things experience a complicated phenomenon called photodegradation, in which their molecular structure goes through a sequence of intricate alterations. Plastics are physically broken down into ever-smaller pieces through a series of complex steps. In several ways, organisms, especially those that filter food, might be negatively impacted by the widespread occurrence of microplastic deposition in the natural environment. These organisms mistake these tiny plastic particles for food and ingest them along with their usual food sources. Increasing our understanding of plastic waste's complex composition relies heavily on the categorisation and classification processes we put in place. Categorising plastic waste based on its polymer composition, recyclability potential, and other distinguishing features is an essential part of plastic waste management. This includes a wide range of plastics, such as polyethylene (PE), polypropylene (PP), polyethylene terephthalate (PET), polystyrene (PS), and polyvinyl chloride (PVC) (Goodship 2007). By dissecting these basic elements, we may learn more about the complex nature of plastic trash and develop better approaches to waste management and environmental protection. Some plastics have a stronger propensity for recycling than others, while others present severe challenges due to their complex compositions or limited commercial appeal, demonstrating a wide range of inequalities in their recyclability. Resin identification codes, consisting of a number value enclosed within a triangle arrangement of arrows, are used to categorise plastic garbage. The aforementioned code is used to identify the kind of polymer and hence the most effective means of recycling it. When compared to complex plastic composites like multi-layer packaging or amalgamated polymers, PET (Polyethylene Terephthalate) and HDPE (High-Density Polyethylene) polymers stand out for their exceptional recycling potential. Because of their useful qualities, these polymers are commonly used to make food and drink containers like bottles. Compared to complex plastic composites, PET and HDPE plastics are more adaptable to recycling and eco-friendly waste management because of this (Singh et al. 2017). Therefore, sorting plastic waste into manageable categories helps us better understand its complex makeup and develop more targeted approaches to waste management and recycling.

2.4.3 Microplastics

Microplastics have recently emerged as a major environmental concern, especially in marine environments, prompting widespread concern about their effects on marine life and ecosystems. Due to their extensive distribution and persistent nature in aquatic settings, these microscopic plastic particles, averaging less than 5 millimetres in size, have gained substantial interest. Microplastics are, primarily in definition, a heterogeneous mixture of very small pieces of plastic that can either be a by-product of the breakdown of bigger plastic items or be produced for specific ends (Laskar and Kumar 2019). It has been hypothesised that billions of microplastics are floating about in the vastness of our seas. These tiny entities have a remarkable tendency to distribute unequally throughout several marine ecosystems, leading to different densities in each of these unique watery environments (Issac and Kandasubramanian 2021). One possible route for the bioaccumulation and biomagnification of hazardous chemicals associated with microplastics is their ingestion by marine species and subsequent incorporation into the food chain. The signs of imbalanced lipid and carbohydrate metabolism and oxidative stress have also been observed in the affected fish livers due to the injection of microplastics (Figure 2.9) (Zolotova et al. 2022). Microplastics also represent a serious risk to filter-feeding species in the marine environment because they clog their filters and make it difficult to obtain water nutrients. Examples of such creatures that are especially susceptible to the harmful impacts of microplastics are mussels and oysters. When these bivalves consume microplastics with their normal diet, it can disrupt their physiological processes by causing internal blockages, reducing feeding efficiency and even killing them (Wright, Thompson, and Galloway 2013).

Understanding the sources, distinguishing qualities, and inherent dangers of microplastics relies heavily on the process of identifying and classifying them. Recent findings have broadened the characteristics used to categorise microplastics, including their size, shape, and polymer composition. Primary microplastics and secondary microplastics are the two most common types of microplastics. The former group includes microbeads, which are used in cosmetics, and pellets, which are widely used in several industrial procedures. The latter group, known as secondary microplastics, develops from the slow breakdown of bigger plastic waste (Laskar and Kumar 2019). The rate of polymer biodegradation for the formation of microplastics increases as particle size decreases and surface area increases (**Figure 2.10(a)**) (Hale et al. 2020). Primary particles are the undisputed kings of microplastics, containing a wide variety of microscopic entities that can have serious effects on our ecosystem. Famous microbeads are among them; they are used as exfoliants and cleaning agents in cosmetics and toiletries. Similarly, we come into contact with textile microfibers that float silently through the air and water, permanently altering these systems. Finally, there are the unassuming pellets, which are used in a variety of industrial contexts despite the fact that their small size belies their capacity to upset natural systems. As a group, these major microplastics pose a significant threat and need our careful consideration as we make our way through the maze of plastic pollution. Macroscopic plastic items, such as plastic bags, bottles, and packaging, degrade and disintegrate over time, releasing secondary microplastics into the environment (Horton and Dixon 2018). These particles can enter the environment in a variety of ways, such as by direct emission, surface runoff, and atmospheric deposition. Microplastics have been found in a wide variety of environments, including the open ocean, freshwater bodies, terrestrial soil, and even, in a startling new finding, microplastics have been found in the air. The wide variety of microplastics might exhibit varied environmental behaviour, with, therefore, varied ecological consequences (Silva et al. 2018).

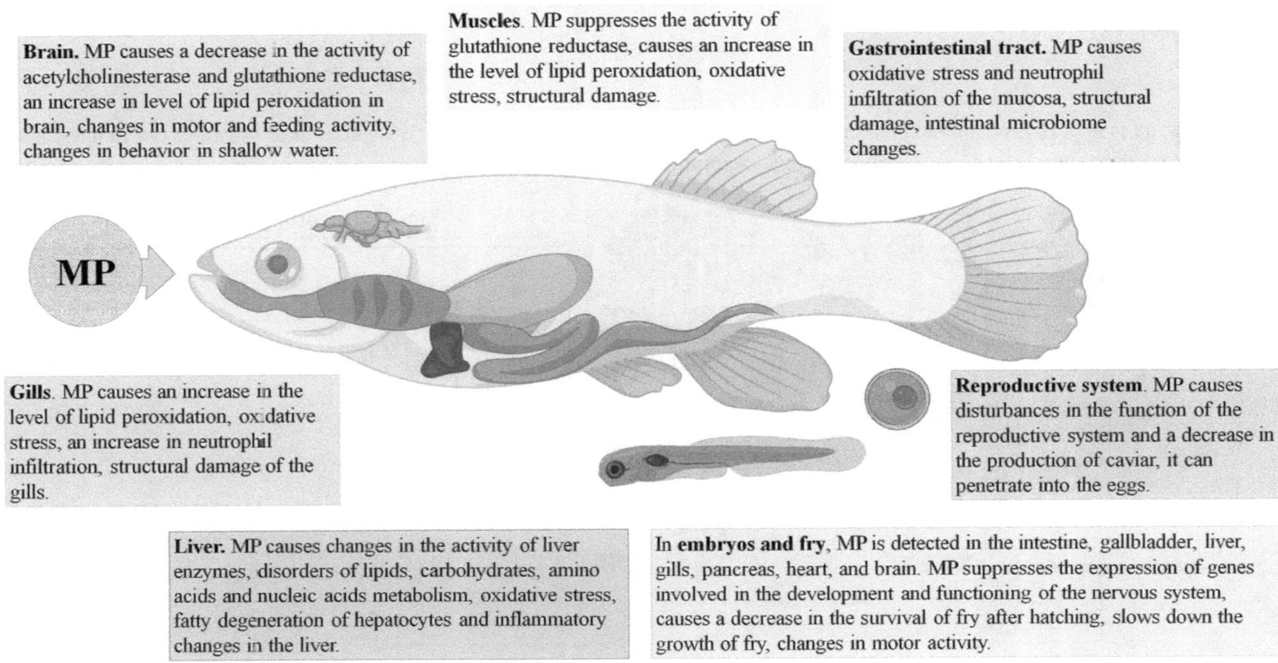

FIGURE 2.9 Harmful effects of microplastics (MP) on marine life. (Zolotova et al. 2022.)

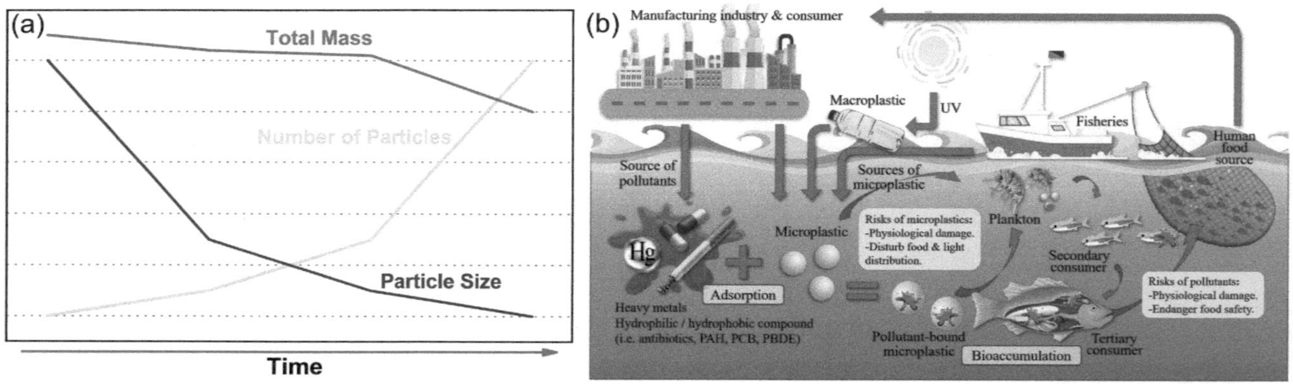

FIGURE 2.10 (a) Relationship between microplastic size, particle number and total mass over time (Hale et al. 2020); (b) An illustration of the pathways of microplastic and its general impact to the marine ecosystem and organisms. (Amelia et al. 2021.)

Microplastics have a wide variety of complex effects on ecosystems, wildlife, and possibly human health. The risks associated with microplastics extend far beyond the field of ecological repercussions and should cause concern for the safety of the general public. Scientific studies have uncovered strong evidence that microplastics may accumulate harmful compounds from their near environment. Substances like persistent organic pollutants (POPs) and heavy metals are included in this category. These compounds can enter the bodies of the species that eat them and persist there, opening the door for them to enter the human food chain through the intake of seafood (Kinigopoulou et al. 2022). Figure 2.10(b) shows an illustration of the pathways of microplastic and its general impact to the marine ecosystem and organisms (Amelia et al. 2021). While research on the effects of microplastics on human health is ongoing, growing concerns have been voiced about the long-term effects of microplastic intake and the existence of dangerous chemicals. It is possible for species at different levels of the food web to ingest microplastics leading to the accumulation and amplification of hazardous chemicals (Yang et al. 2022). Numerous physiological and reproductive effects can stem from the aforementioned occurrence, causing organisms to alter their eating habits and ultimately suffer a decline in their fitness levels. In both aquatic and terrestrial environments, microplastics have also been shown to impair important ecological processes such as nutrient cycle, sediment composition, and biodiversity. Besides, microplastics' ecological repercussions are able to induce disturbances within food webs, alter species relationships, and impact the overarching health and operational dynamics of ecosystems. Furthermore, significant data has emerged detailing the possible effects of microplastics on the immune system and the disruption of endocrine function in animals (Prata et al. 2020). Although to understand the entire breadth and depth of these impacts, more research is required.

2.5 CHEMICAL INDUSTRIAL WASTE

Toxic materials produced by businesses, institutions, and industries engaged in chemical production, use, or research are collectively called "industrial chemical waste." Toxic chemicals, combustible materials, corrosive agents, and reactive molecules are all included in the broad category of industrial chemical waste. As a result of its intrinsic capability for contamination and its extraordinary ability to survive for lengthy periods within the ecosystem, chemical waste poses significant environmental concerns when it is present in our surroundings (Dunne et al. 1990). An exhaustive study by the United Nations Environment Programme (UNEP) has concluded that each year the world produces almost 400 million tonnes of dangerous chemicals. The inevitable by-product of such a staggering rate of production is a mountain of unused chemical waste (Orloff and Falk 2003). An extensive study was done to determine the effect of chemical exposure in the environment on the total burden of disease, and the results were published in the prestigious medical journal Lancet. According to the results of this in-depth analysis, such exposure is a major factor in the estimated 1.6 million annual fatalities (Fuller et al. 2022). The foregoing realisation highlights the critical need for the introduction of efficient chemical waste categorisation and management methods, with the primary aim of protecting human health and ecosystem balance. The environmentally detrimental issues surrounding industrial chemical waste include soil and land contamination, water and air pollution as well as ecosystem disruption. The development and use of effective management strategies and mitigation measures depend critically on understanding the environmental consequences of this category of technological waste.

2.5.1 Environmental and Ecological Consequences

Negative effects of chemical waste on soil composition, productivity, and ecological dynamics result from **soil and land contamination**, which develops as a result of insufficient disposal techniques or the accidental release of chemical waste. Approximately 19% of all occurrences of work-related illnesses and deaths may be traced back to occupational exposure to chemicals, according to estimates published by the highly regarded World Health Organisation (WHO). Understanding the complex mechanisms by which chemical pollutants penetrate the soil matrix is crucial in environmental research (Artiola et al. 2019). This movement of pollutants into the soil, known as leaching, sets off a chain reaction that has far-reaching effects on the soil's physical, chemical, and biological properties. It has been noted that soil contamination has substantial consequences on the complex process of nutrient cycling, consequently obstructing its smooth functioning. Soil microorganisms are essential to the sustainability of a balanced soil ecosystem, and research has shown that the presence of pollutants in soil has a suppressive effect on their activity. Therefore, plants suffer as a result of soil pollution, which has negative effects on plant growth and development. Reduced agricultural output, contaminated groundwater, and harm to species living below ground are all possible results of the aforementioned situation (Misra and Pandey 2005).

Chemical waste contributes to **water contamination** when it is released into water sources through poor disposal methods, unintentional accidents, or runoff from industrial facilities. Water quality, the delicate balance of aquatic life, and the potential threats to human health are all seriously affected by the buildup of toxins within aquatic ecosystems, making this an issue of significant relevance. The presence of chemical pollutants has been shown to build up over time in aquatic settings. The quality of water and the delicate balance of aquatic ecosystems have been shown to suffer when this occurrence continues uncontrolled. As a result, there is a serious threat to human health from consuming water or seafood that has been polluted with these chemicals. Heavy metals and certain organic contaminants are examples of persistent and bioaccumulative chemicals that might cause special concerns. Many different properties, such as solubility, persistence, and bioaccumulation potential, are associated with chemical contaminants. Amazingly, bodies of water may act as reservoirs for these harmful compounds, gathering them over time (Bolong et al. 2009). Toxic compounds in the environment are taken up by creatures living there and eventually make their way up the food chain to the top, a process known as bioaccumulation.

The emission of volatile compounds from chemical waste causes **air pollution**, which has a negative impact on air quality and poses a serious threat to the respiratory health of the human population. Harmful compounds may be released into the atmosphere when chemical waste is burned (Ukaogo, Ewuzie, and Onwuka 2020). This might result from air pollution, including greenhouse gases, particulate matter, and volatile organic compounds (VOCs). VOCs have the unique property of easily evaporating at room temperature. They can contribute to the formation of tropospheric ozone and photochemical smog when exposed to sunlight (Pandey and Yadav 2018). Combustion of chemical waste results in the release of potentially harmful gases such as nitrogen oxides (NO_x), sulphur dioxide (SO_2), and carbon monoxide (CO). Understanding the complex nature of the chemicals and gases released by chemical waste is essential in the field of chemical waste management. When discharged into the air, these compounds may contribute greatly to the widespread problem of air pollution (Kampa and Castanas 2008). This phenomenon has far-reaching consequences since it immediately impacts the quality of the air we breathe, which in turn poses significant threats to the precarious balance of human health. Combustible chemical waste, if incinerated or improperly burned, can also release harmful fumes and tiny solid particles (sometimes known as "particulate matter"). The Bhopal Gas Tragedy of 1984 is a tragic example of the far-reaching effects of improperly storing and disposing of chemical waste. The tragic accident that resulted in the accidental release of poisonous methyl isocyanate gas from a pesticide factory in the Indian city of Bhopal is a mournful witness to the devastating effects of an industrial accident. This tragic occurrence happened not too long ago, and it caused the premature deaths of many people and left others who survived with long-lasting health problems (Broughton 2005). This incident highlighted the need for stringent regulatory measures, careful handling procedures, and vigilant inspection of chemical waste to prevent future disasters.

Industrial chemical waste has the potential to severely disrupt the delicate **equilibrium of ecosystems**, with far-reaching consequences for many different aspects of ecology. Changes in biodiversity, population dynamics, and the stability and efficiency of ecosystems are all indicators of this upheaval. Chemical waste is particularly dangerous because of its toxicity and ability to stay in the environment (Ayangbenro and Babalola 2017). When harmful compounds enter the environment, they set a cascade of reactions with potentially far-reaching effects in motion. Loss of biodiversity, or the variety and number of species found in an environment, is one of the most significant effects. Some organisms may be more vulnerable to the hazardous effects of chemical waste than others, which can decrease species richness. It has been discovered that the buildup of pollutants within food chains has a considerable influence on creatures at various trophic levels (Rajmohan, Chandrasekaran, and Varjani 2020). Therefore, a shift in species composition can have far-reaching consequences across the food web.

2.5.2 Categorisation of Industrial Chemical Waste

Understanding the harmful effects of chemical waste is insufficient to tackle the problem head-on. It requires digging deep into the nuanced world of classifying chemical waste, a topic of critical relevance in environmental research. Proper management and mitigation of chemical waste's ecological and physiological effects necessitate understanding the many classes and distinguishing qualities connected to this waste. Acidity, toxicity, flammability, and reactivity are only a few of the physical and chemical properties that may be used to define a substance's categorisation.

Harmful chemicals, which cover a wide range of compounds, have characteristics that are bad for people and the environment. Chemical waste contains many compounds with undesirable properties, including those that are poisonous, carcinogenic, mutagenic, or otherwise harmful to the environment. The aforementioned chemicals include many different types of harmful molecules, including volatile organic compounds (VOCs), persistent organic pollutants (POPs), and heavy metals, including lead, cadmium, and mercury (Sharma et al. 2014; Misra and Pandey 2005). Chemical wastes that easily catch fire fall under the category of "**flammable materials**." Organic solvents, fossil fuels, and gases are all examples of such materials, suggesting that this type of waste has a high potential to start fires and even explosions. The aforementioned materials have low flash points and can release flammable fumes when exposed to heat or flame (de Dios Caputto et al. 2022). An important property of any given substance is its flash point, defined as the lowest temperature at which its vapours may be released in sufficient quantity to catch fire when exposed to an external ignition source. Inflammability of substances is determined by their flash points, which are determined by standard testing procedures (Vidal et al. 2004). When it comes to preventing accidents and reducing fire risks, the careful management of combustible waste—including its storage, processing, and final disposal—takes centre stage. Many types of chemical waste also contain **corrosive chemicals** like acids and bases. Due to their corrosive properties, these compounds can cause significant damage to both biological tissues and infrastructure. Corrosive waste is a powerful entity in the world of hazardous chemicals due to its amazing reactivity. Corrosion occurs when a substance has a chemical interaction with its environment, causing the wear and tear of numerous components over time. Chemical reactions may occur between corrosive chemicals and a wide range of materials, including metals and living flesh. These compounds can take on acidic or alkaline properties, which determines the scope and nature of their destructive effects. The presence of **reactive compounds**, including many different types of chemicals, in chemical waste might increase the danger it poses. These substances are inherently unstable, explosive, or capable of producing poisonous or hazardous byproducts during chemical reactions. Scientifically speaking, reactive compounds are substances prone to chemical changes under specific conditions. Temperature shifts, pressure shifts, and interactions with new chemical species all fall within this category. These compounds have remarkable reactivity, often undergoing spontaneous combustion or releasing toxic gases when brought into contact with oxygen or water. Recognising the complexity of reactive waste and the need for particular protocols for its storage and disposal is crucial in the field of waste management.

2.6 NUCLEAR WASTE

The word "nuclear waste" refers to any radioactive substance that has been generated as a result of nuclear power plant operations, nuclear weapon factories, or any other use of nuclear technology. Nuclear waste's intrinsic radioactivity and the resulting potential for lasting contamination are major causes for alarm over its environmental effects. It has been proven scientifically that the radiation emitted by nuclear waste has a long half-life, measured in millennia (Ojovan and Steinmetz 2022). This prolonged persistence creates serious threats to natural systems and human health. The half-lives of some radioactive isotopes inside nuclear waste can stretch hundreds of thousands of years or even many millennia. The nuclear disasters at Chernobyl (Medvedev 1986) and Fukushima (Evrard, Laceby, and Nakao 2019) act as striking examples of the far-reaching ecological effects of such catastrophes. Large amounts of radioactive compounds were released into the environment as a result of the aforementioned events, leading to contamination of the ground, water supplies, and air. The harmful effects of these catastrophes have also been shown to last for a long time, with the affected population suffering from long-term health problems.

2.6.1 Effects on Ecology and Environment

The nuclear waste management paradigm requires a deep dive into the complexities of radioactive materials. Ionising radiation emission is a distinctive property of these compounds. This type of radiation has the ability to disrupt the delicate balance of ecosystems and harm living things due to the energy it contains. Nuclear waste contains ionising radiation, the effects of which must be well comprehended. Exposure to such radiation has far-reaching detrimental biological and ecological effects that go far beyond simple physical injury (Santana 2016). Due to the complex nature of the relationship between radioactive elements and the web of life, a thorough analysis of their consequences is required. Complex cellular architecture and dynamic physiological processes make living beings a prime example of this. It is widely established that exposure to ionising radiation may damage DNA, leading to the development of new mutations and an increase in cancer risk. Environmental contamination from nuclear waste can have far-reaching effects on the ecosystem and the delicate balance of biodiversity, in addition to the direct effects on human health discussed above (Political Ecology Research Group 1982). This is because some radioactive isotopes in nuclear waste have half-lives that are hundreds of times longer than those of the elements in the environment. Additionally, we can't lose sight of how sneaky radiation-induced genetic abnormalities are since they keep passing on inherited problems from generation to generation. Exposure to high amounts of radiation from nuclear waste clearly has far-reaching and significant effects on human health (Adeola et al. 2023). Due to the persistent dangers posed by nuclear waste, stringent safety precautions are required to reduce the likelihood of people being exposed to radiation.

2.6.2 Classifying Nuclear Waste

Understanding the properties and decay rates of nuclear waste as well as developing effective management strategies, relies heavily on its categorisation and classification. To effectively deal with the wide variety of nuclear waste types, it is necessary to establish separate methods. Scientists and researchers obtain important insights into the particular properties of nuclear waste by methodical organisation and classifying, allowing them to develop effective strategies for its safe treatment and disposal. This method allows for a thorough comprehension of the many nuclear waste types, their individual decay rates, and the actions necessary to limit the dangers associated with their long-term storage and final disposal (Weber and Roberts 1983). Classifying nuclear waste is a complex task in the field of nuclear research, depending on a wide range of important characteristics and behaviours. These characteristics constitute the basis for categorising this powerful by-product, which includes its radioactivity, heat production, and lifespan. The degree of radioactivity present in nuclear waste is a crucial factor in its categorisation. To what extent the waste emits ionising radiation is measured by this metric. This phenomenon results from the unstable atomic nuclei in nuclear waste. Experts can identify the possible dangers of nuclear waste through radioactivity assessment, allowing for safer waste management (IAEA 1994). The amount of heat produced is also very important when deciding how to categorise nuclear waste. The results of the classification process all heavily influence storage, treatment, and disposal strategies.

The International Atomic Energy Agency (IAEA) nuclear waste classification system (Figure 2.11), used to categorise nuclear waste in terms of activity content,

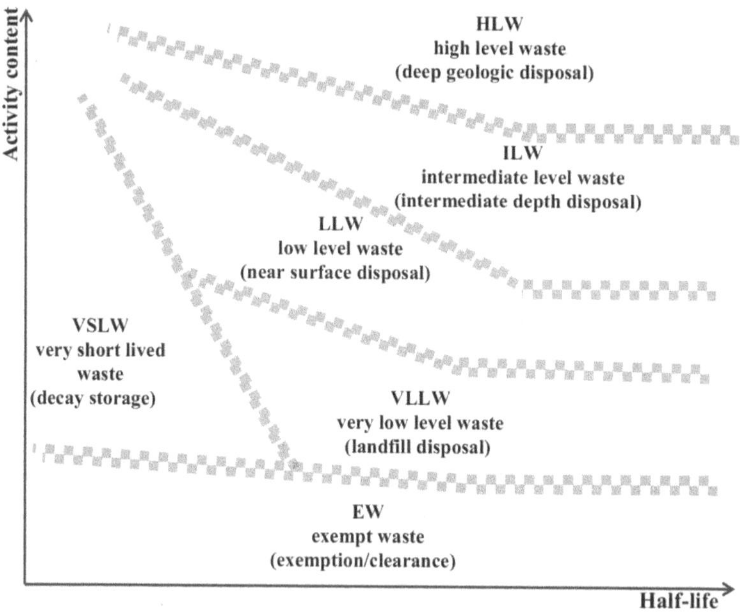

FIGURE 2.11 Schematic of IAEA radioactive waste classification scheme. The term "activity content" covers activity concentration and total activity. (IAEA 1994.)

is one of the most objective and prominent methods (IAEA 1994). Low-level waste (LLW) is defined as having low radioactive activity and includes many different types of items, such as but not limited to protective clothing, tools, and components used in nuclear reactor operations. It is important to note that LLW possesses a unique trait allowing for its secure disposal in repositories close to the Earth's surface. The increased radioactivity of intermediate-level waste (ILW) defines it as a subset of radioactive waste. Materials like resins, filters, and even some parts from nuclear reactors fall within this umbrella. Both spent nuclear fuel and the remaining by-products of reprocessing spent fuel fall under the umbrella term "high-level waste" (HLW) due to their high radioactivity levels. The term "spent nuclear fuel" is used to describe the nuclear fuel that is removed from reactors once it can no longer power a self-sustaining chain reaction (IAEA 1994). HLW must be disposed of in deep geological repositories where it will remain isolated for an extended period of time. Each year, the world's nuclear power reactors generate around 2,500 metric tonnes of HLW, according to the IAEA (Agency 2021). This interesting data highlight the worldwide significance of nuclear power generation. Such an essential categorisation framework is fundamental to properly managing and disposing of radioactive waste, which in turn protects humans and the environment.

2.7 SUMMARY

This chapter thoroughly investigates the numerous categories of technological waste and has highlighted their negative effects on the natural environment and the well-being of humans. The discussion on mining and metal extraction waste, electronic waste (e-waste), chemical waste, and radioactive waste has shown a complex web of tremendous impediments that require society's immediate attention. The harmful consequences of e-waste are brought to light by the ceaseless advancement of technology and the concomitant increase in the use of electronic gadgets. Their incorrect disposal and inadequate recycling processes result in the release of hazardous compounds like lead, mercury, and cadmium into the environment. Furthermore, the negative impact of waste created by various sectors relating to metal extraction and beneficiation processes must be acknowledged, as it considerably contributes to both environmental degradation and potential health dangers. The release of hazardous compounds, such as heavy metals, solvents, and pesticides, has the potential to contaminate water reservoirs, terrestrial substrates, and even the food that we consume. The link between long-term exposure to these carcinogenic substances and a variety of ailments is well-established, encompassing a wide range of ailments ranging from respiratory impairments and compromised organ functionality to developmental anomalies and, in the most severe cases, malignant neoplasms. To address these serious issues, a comprehensive and diversified plan that includes the participation and concentrated efforts of governmental entities, industry sectors, and individual stakeholders is required. The deployment of improved waste management systems, including proficient recycling procedures and proper disposal strategies, is critical in limiting the negative consequences of technology waste. The promotion and growth of sustainable technologies that emphasise durability, recyclability, and environmental compatibility can act as a catalyst in reducing waste generation, which will be the salient points for discussion in the forthcoming chapters. In conclusion, a thorough understanding of the various types of technological waste and their negative consequences represents a critical step in the development of methodologies aimed at waste reduction, streamlined recycling practices, and the protection of our ecological environment and overall well-being.

REFERENCES

Abdelbasir, Sabah M., Saad S. M. Hassan, Ayman H. Kamel, and Rania Seif El-Nasr. 2018. Status of electronic waste recycling techniques: A review. *Environmental Science and Pollution Research* 25 (17):16533–16547.

Adeola, A. O., K. O. Iwuozor, K. G. Akpomie, et al. 2023. Advances in the management of radioactive wastes and radionuclide contamination in environmental compartments: A review. *Environmental Geochemistry and Health* 45 (6):2663–2689.

Akcil, Ata, and Soner Koldas. 2006. Acid Mine Drainage (AMD): Causes, treatment and case studies. *Journal of Cleaner Production* 14 (12):1139–1145.

Alabi, Okunola A., Yetunde M. Adeoluwa, Xia Huo, Xijin Xu, and Adekunle A. Bakare. 2021. Environmental contamination and public health effects of electronic waste: An overview. *Journal of Environmental Health Science and Engineering* 19 (1):1209–1227.

Ali, Sameh Samir, Tamer Elsamahy, Eleni Koutra, et al. 2021. Degradation of conventional plastic wastes in the environment: A review on current status of knowledge and future perspectives of disposal. *Science of the Total Environment* 771:144719.

Amelia, Tan Suet May, Wan Mohd Afiq Wan Mohd Khalik, Meng Chuan Ong, Yi Ta Shao, Hui-Juan Pan, and Kesaven Bhubalan. 2021. Marine microplastics as vectors of major ocean pollutants and its hazards to the marine ecosystem and humans. *Progress in Earth and Planetary Science* 8 (1):12.

Andrady, A. L., and M. A. Neal. 2009. Applications and societal benefits of plastics. *Philosophical Transactions of the Royal Society of London* 364 (1526):1977–1984.

Artiola, J. F., J. L. Walworth, S. A. Musil, and M. A. Crimmins. 2019. Chapter 14 – Soil and land pollution. In *Environmental and Pollution Science (Third Edition)*, edited by M. L. Brusseau, I. L. Pepper, and C. P. Gerba: Academic Press.

Awasthi, Abhishek Kumar, Xianlai Zeng, and Jinhui Li. 2016. Environmental pollution of electronic waste recycling in India: A critical review. *Environmental Pollution* 211:259–270.

Ayangbenro, Ayansina Segun, and Olubukola Oluranti Babalola. 2017. A new strategy for heavy metal polluted environments: A review of microbial biosorbents. *International Journal of Environmental Research and Public Health* 14 (1):94.

Bajpai, Anurag, Partha Kumbhakar, Chandra Sekhar Tiwary, and Krishanu Biswas. 2021. Conducting graphene synthesis from electronic waste. *ACS Sustainable Chemistry & Engineering* 9 (42):14090–14100.

Beula, D., and M. Sureshkumar. 2021. A review on the toxic E-waste killing health and environment – Today's global scenario. *Materials Today: Proceedings* 47:2168–2174.

Bolong, N., A. F. Ismail, M. R. Salim, and T. Matsuura. 2009. A review of the effects of emerging contaminants in wastewater and options for their removal. *Desalination* 239 (1):229–246.

Broughton, Edward. 2005. The Bhopal disaster and its aftermath: A review. *Environmental Health* 4 (1):6.

Brown, Valerie J. 2003. REACHing for chemical safety. *Environmental Health Perspectives* 111 (14):A766–A769.

Chapman, Neil A. 2006. Geological disposal of radioactive wastes–Concept, status and trends. *Journal of Iberian Geology* 32 (1):7–14.

Croff, A. G., A. A. Richmond, and J. P. Williams. 1995. Historical literature review on waste classification and categorization Oak Ridge National Lab. (ORNL), Oak Ridge, TN, United States.

Das, B., S. Prakash, P. S. R. Reddy, and V. N. Misra. 2007. An overview of utilization of slag and sludge from steel industries. *Resources, Conservation and Recycling* 50 (1):40–57.

Daum, Kurt, Justin Stoler, and Richard J. Grant. 2017. Toward a more sustainable trajectory for E-waste policy: A review of a decade of E-waste research in Accra, Ghana. *International Journal of Environmental Research and Public Health* 14 (2):135.

de Dios Caputto, María Dolores, Rodrigo Navarro, Juan López Valentín, and Ángel Marcos-Fernández. 2022. Chemical upcycling of poly(ethylene terephthalate) waste: Moving to a circular model. *Journal of Polymer Science* 60 (24):3269–3283.

Decharat, S., and P. Kiddee. 2020. Health problems among workers who recycle electronic waste in Southern Thailand. *Osong Public Health and Research Perspectives* 11 (1):34–43.

Ding, Peng, Quan Jun Liu, and Wen Hao Pang. 2013. A review of manganese ore beneficiation situation and development. *Applied Mechanics and Materials* 380–384:4431–4433.

Dunne, Michael, Jan Lawton, Beverley Raphael, and Paul Burnett. 1990. The health effects of chemical waste in an urban community. *Medical Journal of Australia* 152 (11):592–597.

Dutta, Joystu, and Moharana Choudhury. 2018. Plastic pollution: A global problem from a local perspective. *Journal of Waste Management & Xenobiotics* 1 (1):000102.

Eto, Komyo, Masumi Marumoto, and Motohiro Takeya. 2010. The pathology of methylmercury poisoning (Minamata disease). *Neuropathology* 30 (5):471–479.

Evrard, O., J. P. Laceby, and A. Nakao. 2019. Effectiveness of landscape decontamination following the Fukushima nuclear accident: A review. *Soil* 5 (2):333–350.

Forti, V., Baldé, C. P., Kuehr, R., & Bel, G. (2020). The global e-waste monitor 2020. *United Nations University (UNU), International Telecommunication Union (ITU) & International Solid Waste Association (ISWA), Bonn/Geneva/Rotterdam, 120.*

Fuller, Richard, Philip J. Landrigan, Kalpana Balakrishnan, et al. 2022. Pollution and health: A progress update. *The Lancet Planetary Health* 6 (6):e535–e547.

Galappaththi, Himan K. A., and Inoka Suraweera. 2020. Risk of mercury exposure during childhood: A review of Sri Lankan situation. *Reviews on Environmental Health* 35 (3):229–232.

Goodship, Vannessa. 2007. Plastic recycling. *Science Progress* 90 (4):245–268.

Halden, Rolf U. 2010. Plastics and health risks. *Annual Review of Public Health* 31 (1):179–194.

Hale, Robert C., Meredith E. Seeley, Mark J. La Guardia, Lei Mai, and Eddy Y. Zeng. 2020. A global perspective on microplastics. *Journal of Geophysical Research: Oceans* 125 (1):e2018JC014719.

Heacock, Michelle, Bain Kelly Carol, Ansong Asante Kwadwo, et al. 2016. E-waste and harm to vulnerable populations: A growing global problem. *Environmental Health Perspectives* 124 (5):550–555.

Hees, Timo, Fan Zhong, Markus Stürzel, and Rolf Mülhaupt. 2019. Tailoring hydrocarbon polymers and all-hydrocarbon composites for circular economy. *Macromolecular Rapid Communications* 40 (1):1800608.

Horton, Alice A., and Simon J. Dixon. 2018. Microplastics: An introduction to environmental transport processes. *WIREs Water* 5 (2):e1268.

Hosler, D., S. L. Burkett, and M. J. Tarkanian. 1999. Prehistoric polymers: Rubber processing in ancient mesoamerica. *Science* 284 (5422):1988–1991.

Houot, R. 1983. Beneficiation of iron ore by flotation — Review of industrial and potential applications. *International Journal of Mineral Processing* 10 (3):183–204.

Huang, W., Shi, X., & Wu, K. (2021). Human Body Burden of Heavy Metals and Health Consequences of Pb Exposure in Guiyu, an E-Waste Recycling Town in China. *International Journal of Environmental Research and Public Health, 18*(23), 12428.

International Atomic Energy Agency, (1994). Classification of Radioactive Waste Safety Guide, IAEA Safety Series No. 111-G-1.1, IAEA, Vienna.

International Atomic Energy Agency. 2021. *IAEA Increases Projections for Nuclear Power Use in 2050.* www.iaea.org/newscenter/pressreleases/iaea-increases-projections-for-nuclear-power-use-in-2050

Issac, Merlin N., and Balasubramanian Kandasubramanian. 2021. Effect of microplastics in water and aquatic systems. *Environmental Science and Pollution Research* 28 (16):19544–19562.

Jain, Manish Kumar, and Arpita Das. 2017. Impact of mine waste leachates on aquatic environment: A review. *Current Pollution Reports* 3 (1):31–37.

Jamieson, Heather E., Stephen R. Walker, and Michael B. Parsons. 2015. Mineralogical characterization of mine waste. *Applied Geochemistry* 57:85–105.

Kampa, Marilena, and Elias Castanas. 2008. Human health effects of air pollution. *Environmental Pollution* 151 (2):362–367.

Kiddee, Peeranart, Ravi Naidu, and Ming H. Wong. 2013. Electronic waste management approaches: An overview. *Waste Management* 33 (5):1237–1250.

Kim, Jihye, and Gisele Azimi. 2021. Valorization of electric arc furnace slag via carbothermic reduction followed by acid baking–Water leaching. *Resources, Conservation and Recycling* 173:105710.

Kinigopoulou, Vasiliki, Ioannis Pashalidis, Dimitrios Kalderis, and Ioannis Anastopoulos. 2022. Microplastics as carriers of inorganic and organic contaminants in the environment: A review of recent progress. *Journal of Molecular Liquids* 350:118580.

Kossoff, D., W. E. Dubbin, M. Alfredsson, S. J. Edwards, M. G. Macklin, and K. A. Hudson-Edwards. 2014. Mine tailings dams: Characteristics, failure, environmental impacts, and remediation. *Applied Geochemistry* 51:229–245.

Kumar, Amit, Maria Holuszko, and Denise Crocce Romano Espinosa. 2017. E-waste: An overview on generation, collection, legislation and recycling practices. *Resources, Conservation and Recycling* 122:32–42.

Laidlaw, Mark A. S., Gabriel M. Filippelli, Sally Brown, et al. 2017. Case studies and evidence-based approaches to addressing urban soil lead contamination. *Applied Geochemistry* 83:14–30.

Laskar, Nirban, and Upendra Kumar. 2019. Plastics and microplastics: A threat to environment. *Environmental Technology & Innovation* 14:100352.

Ledin, M., and K. Pedersen. 1996. The environmental impact of mine wastes — Roles of microorganisms and their significance in treatment of mine wastes. *Earth-Science Reviews* 41 (1):67–108.

Li, Liancheng, Jian Zuo, Xiaoguang Duan, Shaobin Wang, Kunsheng Hu, and Ruidong Chang. 2021. Impacts and mitigation measures of plastic waste: A critical review. *Environmental Impact Assessment Review* 90:106642.

Li, Weila, and Varenyam Achal. 2020. Environmental and health impacts due to e-waste disposal in China – A review. *Science of the Total Environment* 737:139745.

Lottermoser, Bernd G. 2010. Sulfidic mine wastes. In *Mine Wastes: Characterization, Treatment and Environmental Impacts*, edited by B. Lottermoser. Berlin, Heidelberg: Springer.

Lottermoser, Bernd G. 2010. Tailings. In *Mine Wastes: Characterization, Treatment and Environmental Impacts*, edited by B. Lottermoser. Berlin, Heidelberg: Springer.

Ma, Yulong, William A. Stubbings, Reginald Cline-Cole, and Stuart Harrad. 2021. Human exposure to halogenated and organophosphate flame retardants through informal e-waste handling activities – A critical review. *Environmental Pollution* 268:115727.

Medvedev, Zhores A. 1986. Ecological aspects of the chernobyl nuclear plant disaster. *Trends in Ecology & Evolution* 1 (1):23–25.

Min, Dong Joon, and Fumitaka Tsukihashi. 2017. Recent advances in understanding physical properties of metallurgical slags. *Metals and Materials International* 23 (1):1–19.

Misra N. R., S. Kumar and A. Jain. 2021 "A Review on E-waste: Fostering the Need for Green Electronics," *International Conference on Computing, Communication, and Intelligent Systems (ICCCIS)*, Greater Noida, India, 2021, 1032-1036

Misra, Virendra, and S. D. Pandey. 2005. Hazardous waste, impact on health and environment for development of better waste management strategies in future in India. *Environment International* 31 (3):417–431.

Oberle, B., Bringezu, S., Hatfield-Dodds, S., Hellweg, S., Schandl, H., & Clement, J. (2019). *Global resources outlook: 2019*. International Resource Panel, United Nations Envio, Paris, France.

Ojovan, Michael I., and Hans J. Steinmetz. 2022. Approaches to disposal of nuclear waste. *Energies* 15 (20):7804.

Orloff, Kenneth, and Henry Falk. 2003. An international perspective on hazardous waste practices. *International Journal of Hygiene and Environmental Health* 206 (4):291–302.

Pandey P, Yadav R. A Review on Volatile Organic Compounds (VOCs) as Environmental Pollutants: Fate and Distribution. IJPE [Internet]. 2018Jul.31 [cited 2025Jan.11]; 4(02):14–26

Perkins, Devin N., Marie-Noel Brune Drisse, Tapiwa Nxele, and Peter D. Sly. 2014. E-waste: A global hazard. *Annals of Global Health* 80 (4):286–295.

Piatak, Nadine M., Michael B. Parsons, and Robert R. Seal. 2015. Characteristics and environmental aspects of slag: A review. *Applied Geochemistry* 57:236–266.

Political Ecology Research Group, Oxford. :*The Impact of Nuclear Waste Disposals to the Marine Environment*. Oxford: Political Ecology Research Group.

Potysz, Anna, Eric D. van Hullebusch, and Jakub Kierczak. 2018. Perspectives regarding the use of metallurgical slags as secondary metal resources – A review of bioleaching approaches. *Journal of Environmental Management* 219:138–152.

Prata, Joana Correia, João P. da Costa, Isabel Lopes, Armando C. Duarte, and Teresa Rocha-Santos. 2020. Environmental exposure to microplastics: An overview on possible human health effects. *Science of the Total Environment* 702:134455.

Rajmohan, K. S., Ramya Chandrasekaran, and Sunita Varjani. 2020. A review on occurrence of pesticides in environment and current technologies for their remediation and management. *Indian Journal of Microbiology* 60 (2):125–138.

Rautela, Rahul, Shashi Arya, Shilpa Vishwakarma, Jechan Lee, Ki-Hyun Kim, and Sunil Kumar. 2021. E-waste management and its effects on the environment and human health. *Science of the Total Environment* 773:145623.

Ruan, Yaoyang, Dongsheng He, and Ruan Chi. 2019. Review on Beneficiation Techniques and Reagents Used for Phosphate Ores. Minerals, 9 (4):253.

Salgado-Almeida, Bryan, Daniel A. Falquez-Torres, Paola L. Romero-Crespo, Priscila E. Valverde-Armas, Fredy Guzmán-Martínez, and Samantha Jiménez-Oyola. 2022. Risk Assessment of Mining Environmental Liabilities for Their Categorization and Prioritization in Gold-Mining Areas of Ecuador. *Sustainability*, 14 (10):6089.

Santana, Luis Paulo. 2016. Management of radioactive waste: A review. *Proceedings of the International Academy of Ecology & Environmental Sciences* 6 (2):38.

Sharma, Brij Mohan, Girija K. Bharat, Shresth Tayal, Luca Nizzetto, and Thorjørn Larssen. 2014. The legal framework to manage chemical pollution in India and the lesson from the Persistent Organic Pollutants (POPs). *Science of the Total Environment* 490:733–747.

Sharma, Nidhi, Anurag Bajpai, Pradeep Kumar Yadav, et al. 2020. Green route for beneficiation of metallic materials from electronic waste for selective reduction of CO2. *ACS Sustainable Chemistry & Engineering* 8 (32):12142–12150.

Shengo, L. M. 2021. Review of practices in the managements of mineral wastes: The case of waste rocks and mine tailings. *Water, Air, & Soil Pollution* 232 (7):273.

Shittu, Olanrewaju S., Ian D. Williams, and Peter J. Shaw. 2021. Global E-waste management: Can WEEE make a difference? A review of e-waste trends, legislation, contemporary issues and future challenges. *Waste Management* 120:549–563.

Silva, Ana B., Ana S. Bastos, Celine I. L. Justino, João P. da Costa, Armando C. Duarte, and Teresa A. P. Rocha-Santos. 2018. Microplastics in the environment: Challenges in analytical chemistry – A review. *Analytica Chimica Acta* 1017:1–19.

Singh, Narinder, David Hui, Rupinder Singh, I. P. S. Ahuja, Luciano Feo, and Fernando Fraternali. 2017. Recycling of plastic solid waste: A state of art review and future applications. *Composites Part B: Engineering* 115:409–422.

Storey, Andrew W., and Markson Yarrao. 2008. Chapter 17 Development of Aquatic Food Web Models for the Fly River, Papua New Guinea, and their Application in Assessing Impacts of the Ok Tedi Mine. In *Developments in Earth and Environmental Sciences*, edited by B. Bolton: Elsevier, Volume 9, 575–615.

Terazono, Atsushi, Shinsuke Murakami, Naoya Abe, et al. 2006. Current status and research on E-waste issues in Asia. *Journal of Material Cycles and Waste Management* 8 (1):1–12.

Thushari, Gajahin Gamage Nadeeka, and Jayan Duminda Mahesh Senevirathna. 2020. Plastic pollution in the marine environment. *Heliyon* 6 (8), e04709.

Ukaogo, Prince O., Ugochukwu Ewuzie, and Chibuzo V. Onwuka. 2020. 21 – Environmental pollution: causes, effects, and the remedies. In *Microorganisms for Sustainable Environment and Health*, edited by P. Chowdhary, A. Raj, D. Verma, and Y. Akhter: Elsevier, 419–429.

Verma, Rinku, K. S. Vinoda, M. Papireddy, and A. N. S. Gowda. 2016. Toxic pollutants from plastic waste- A review. *Procedia Environmental Sciences* 35:701–708.

Vidal, M, WJ Rogers, JC Holste, and MS Mannan. 2004. A review of estimation methods for flash points and flammability limits. *Process Safety Progress* 23 (1):47–55.

Walker, Tony R., and Lexi Fequet. 2023. Current trends of unsustainable plastic production and micro(nano)plastic pollution. *TrAC Trends in Analytical Chemistry* 160:116984.

Weber, William J., and Frank P. Roberts. 1983. A review of radiation effects in solid nuclear waste forms. *Nuclear Technology* 60 (2):178–198.

Whitworth, Anne J., James Vaughan, Gordon Southam, et al. 2022. Review on metal extraction technologies suitable for critical metal recovery from mining and processing wastes. *Minerals Engineering* 182:107537.

Wright, Stephanie L., Richard C. Thompson, and Tamara S. Galloway. 2013. The physical impacts of microplastics on marine organisms: A review. *Environmental Pollution* 178:483–492.

Yang, Wenjie, Nahar Jannatun, Yanqiao Zeng, et al. 2022. Impacts of microplastics on immunity. 4. 956885.

Yohannessen, K., D. Pinto-Galleguillos, D. Parra-Giordano, et al. 2019. Health assessment of electronic waste workers in Chile: Participant characterization. *International Journal of Environmental Research Public Health* 16 (3):386.

Zeng, Xiang, Xia Huo, Xijin Xu, Dongling Liu, and Weidong Wu. 2020. E-waste lead exposure and children's health in China. *Science of the Total Environment* 734:139286.

Zolotova, Natalia, Anna Kosyreva, Dzhuliia Dzhalilova, Nikolai Fokichev, and Olga Makarova. 2022. Harmful effects of the microplastic pollution on animal health: A literature review. *PeerJ* 10:e13503.

3 Technological Waste Management

A Brief Overview

Anurag Bajpai

3.1 INTRODUCTION

Within the continuous trajectory of human advancement, an often disregarded consequence is the escalating influx of technological waste. The relentless pursuit of innovation and the unquenchable need for consumer products and electronics have resulted in the widespread presence of abandoned devices, the buildup of industrial waste, and the pervasive use of plastics and chemicals that characterise contemporary society. The generation of technological waste is an inescapable consequence of the swift advancement of our digital era, and it poses a complex and wide-ranging problem that surpasses geographical, disciplinary, and industrial boundaries. Chapter 2 details the various types of technological waste and their harmful effects on humans and ecosystems. However, this dilemma presents a significant potential to utilise science and technology to not only reduce the negative impacts of waste but also to convert it into valuable resources for the development of innovative materials. Mining and metal extraction operations form the fundamental pillars around which our industrial society is built. These operations result in the extraction of essential metals as well as significant quantities of waste materials, including tailings, slag, and mineral leftovers. Below the outside layer of this refuse, there are unexplored reservoirs of important substances, such as rare earth elements, which are vital for cutting-edge technology. In this context, the realms of science and innovation intersect to derive value from materials or resources that were previously deemed to have no worth. Further, the prevalence of electronic gadgets, ranging from cell phones to computers, characterises the contemporary era of digital technology. However, with the rapid advancement of technology, these gadgets quickly become outdated, leading to the emergence of electronic waste. The substantial quantities of abandoned electronic devices not only give rise to environmental apprehensions but also contain valuable resources, including precious metals, rare earth elements, and crucial components essential for the advancement of future technology (Islam et al. 2020). Plastics, known for their versatility and widespread presence, have significantly transformed several facets of our everyday existence. However, the ongoing existence of plastics in both terrestrial and marine environments has sparked a worldwide environmental disaster (Adam et al. 2020). The prevalence of plastic waste and microplastics has prompted scientific innovation aimed at mitigating their environmental consequences

DOI: 10.1201/9781003619680-3

and exploring the possibility of transforming plastics into new materials. The proper handling and disposal of hazardous chemical and radioactive industrial waste holds significance in ensuring the safety of both the environment and humans. The management of these waste streams necessitates the utilisation of specialist materials that possess the ability to endure severe circumstances, ranging from the presence of corrosive chemicals to exposure to ionising radiation (Kumar and Sharma 2019). Scientific advancements in this field are centred on the development of sophisticated materials designed at the nanoscale, where even little modifications at the molecular level can have significant implications.

The common thread that binds these many manifestations of technological waste is their inherent capacity for transformation. The prevailing perspective on waste, which traditionally regarded it as a challenge for control and containment, is undergoing a transformation towards a paradigm that perceives waste as a valuable resource capable of providing raw materials and fostering creativity. The aforementioned transition encompasses not just philosophical aspects but also finds its foundations in the domains of science, engineering, and environmental stewardship. The present chapter provides a thorough examination of the intricate terrain of technical waste management. In this chapter, we undertake a scientific exploration that delves into the fundamental aspects of this transformative thinking. It explores the methodologies, technological advancements, and scientific concepts that facilitate the conversion of waste into advanced materials. Additionally, the discussion emphasises the need to integrate scientific research and innovation within the framework of legal and regulatory requirements. Throughout the chapter, a deep understanding and recognition of the multidisciplinary aspects involved in waste management are deliberated, as well as the crucial significance of scientific approaches in tackling this intricate global issue.

3.2 TECHNOLOGICAL WASTE AS RESOURCES FOR ADVANCED MATERIALS

In contemporary civilisation, the issue of effectively managing technological waste has become a significant concern. However, there is a potential solution around dealing with different technological waste streams that holds promise: the conversion of waste materials into useful resources (Demirbas 2011). The conversion of technological waste into valuable resources signifies a fundamental change in societal perspectives and approaches towards its management. This section explores the possibilities of employing different types of technological waste as antecedents for the advancement and development of novel and advanced materials and technologies.

3.2.1 Waste from Metal Extraction and Mining: A Treasure Trove of Raw Materials

The utilisation of mine tailings, which are commonly perceived as by-products, might be reenvisioned as valuable reservoirs of construction materials (Martins et al. 2021). Geopolymerisation is a chemical process wherein aluminosilicate minerals are activated through the use of alkaline solutions, resulting in the transformation of tailings into geopolymers with qualities similar to cement (Xiaolong et al. 2021). Geopolymers have the potential to function as viable substitutes for conventional construction materials, therefore mitigating the need for resource-intensive substances such as cement and facilitating waste reduction. Mining waste is also found to include small quantities of important essential metals such as cobalt, lithium, and rare-earth elements (Tkaczyk et al. 2018). The utilisation of novel extraction techniques, such as bioleaching employing specialised microbes, facilitates the retrieval of these metallic elements from residual streams. The metals that have been reclaimed can subsequently be used in the manufacturing process of sophisticated technology, such as lithium-ion batteries utilised in electric vehicles and storage systems for renewable energy (Valenta et al. 2023).

3.2.2 Electronic Waste: A Source of Precious Elements

Electronic waste serves as a substantial repository of rare-earth elements (REEs), which are necessary for the functioning of electronic devices and the advancement of environmentally friendly technology. Hydrometallurgical methodologies, such as solvent extraction and ion exchange, exhibit the capability to retrieve REEs from electronic waste selectively (Ambaye et al. 2020). These components are utilised in many applications, such as permanent magnets, catalysts, and efficient lighting systems, hence playing a significant role in the progression of sustainable technology. Moreover, it is worth noting that discarded electronics also include valuable metals such as gold, silver, and palladium within their printed circuit boards (PCBs) (Sethurajan et al. 2019). Metals can be extracted *via* pyrometallurgical techniques that include the controlled heat degradation of the materials. **Figure 3.1** shows the

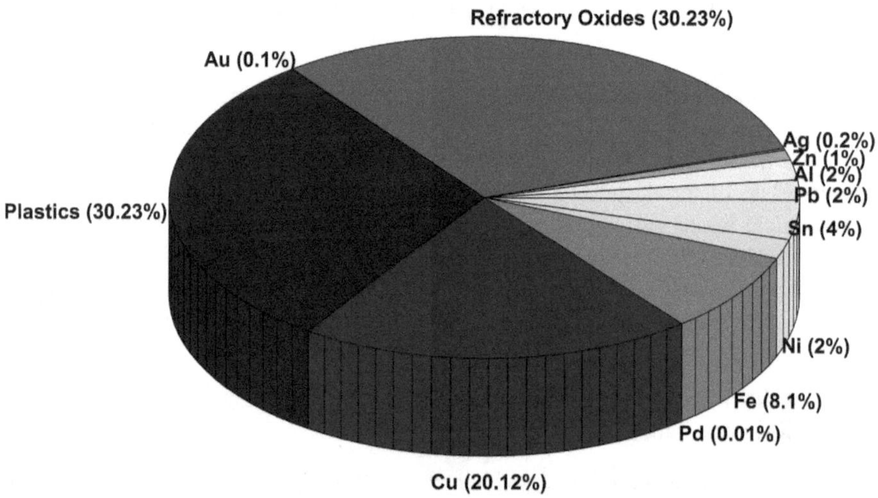

FIGURE 3.1 Average composition of e-waste, showing the proportions of various precous metals, and other materials. (Jaibee et al. 2015.)

average composition of materials (metals and non-metals) in e-waste (Jaibee et al. 2015). The utilisation of retrieved metals in the production of electronic components serves to decrease the dependence on primary resources and mitigate the ecological consequences associated with metal extraction. Nanotechnology also presents innovative methodologies for the extraction of valuable components from electronic waste. Nanoscale particles, such as nanocatalysts, have the ability to extract valuable metals from intricate electronic waste matrices selectively. Moreover, the utilisation of nanomaterials has the potential to augment the efficiency of electrical equipment, hence mitigating the need for limited and ecologically taxing resources (Debnath et al. 2023).

3.2.3 Plastic Waste: Catalysing Innovation in Material Science

The issue of plastic waste offers a potential avenue for upcycling, which involves the conversion of waste materials into goods of greater worth. Chemical recycling techniques, such as pyrolysis and catalytic depolymerisation, facilitate the decomposition of plastic materials into their individual monomers or intermediate compounds (Tian et al. 2023). These materials can serve as feedstocks for the fabrication of new plastics, so establishing a closed-loop system that mitigates the reliance on fossil fuels in the production of plastics. **Figure 3.2** schematically illustrates the bioupcycling technology for sustainable plastic waste management to produce new plastic products

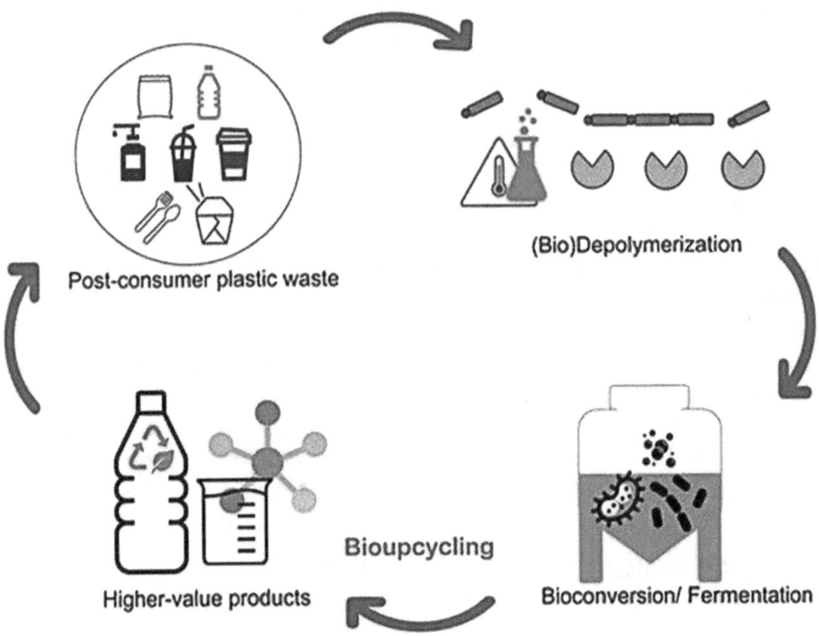

FIGURE 3.2 Bioupcycling technology as a potential solution for sustainable plastic waste management. (Lomwongsopon and Varrone 2022.)

(Lomwongsopon and Varrone 2022). Furthermore, it should be noted that plastic waste, when appropriately treated, has the potential to be utilised as a reinforcing agent in composite materials. One example of such application is the utilisation of shredded plastic in the reinforcement of concrete, which serves to improve its mechanical characteristics (Siddique, Khatib, and Kaur 2008). By effectively redirecting plastic waste away from landfills and incineration, this particular strategy not only aids in the reduction of waste but also enhances the overall performance of building materials. Researchers have successfully increased the mechanical characteristics of materials by integrating shredded plastic waste into a polymer matrix. Nanocomposites have been widely utilised across several industries, ranging from automobile production to aerospace engineering, due to their ability to address waste reduction and enhance material performance simultaneously (Bahadur and Sonal 2016). Bioplastics, which are produced from waste materials such as food scraps and agricultural leftovers, present a viable and environmentally friendly substitute for traditional plastics. Microbial fermentation techniques have the capability to transform organic waste into biopolymers that exhibit comparable characteristics to plastics derived from petroleum. This strategy addresses two concurrent concerns, namely waste management and plastic pollution (Koutinas et al. 2014). Moreover, the advent of additive manufacturing, also known as 3D printing, is significantly transforming several sectors by facilitating the fabrication of highly accurate and personalised items. The conversion of plastic waste into feedstocks that are suitable for 3D printing enables the production of a diverse array of items, spanning from prosthetic devices to architectural elements (Mikula et al. 2021). This technology facilitates a closed-loop system in production, hence mitigating the demand for newly produced plastics.

3.2.4 Industrial Chemical Waste: Unlocking Circular Chemistry

Industrial chemical waste frequently consists of several chemicals that have the potential to be reclaimed and subsequently reintegrated into production processes. Membrane separation techniques, such as pervaporation and ultrafiltration, are employed to effectively and selectively remove valuable solvents and compounds from waste streams (Zaidi, Mauritz, and Hassan 2019). The compounds that have been recovered can serve as viable alternatives to virgin resources in a range of applications, hence promoting the adoption of a circular economy strategy. Chemical waste generated by industrial operations may also function as a valuable fuel for chemical reactions. The notion of "waste-to-chemicals" pertains to the transformation of waste compounds into chemicals of greater value by means of inventive catalytic processes. This methodology not only mitigates the formation of waste but also mitigates the environmental ramifications associated with chemical manufacture.

3.3 TECHNOLOGICAL WASTE MANAGEMENT HIERARCHY: REDUCE, REUSE, RECYCLE, RECOVER, AND DISPOSAL

While the preceding discussion highlights the potential of using different technological waste streams as precursors to developing advanced materials and strategies,

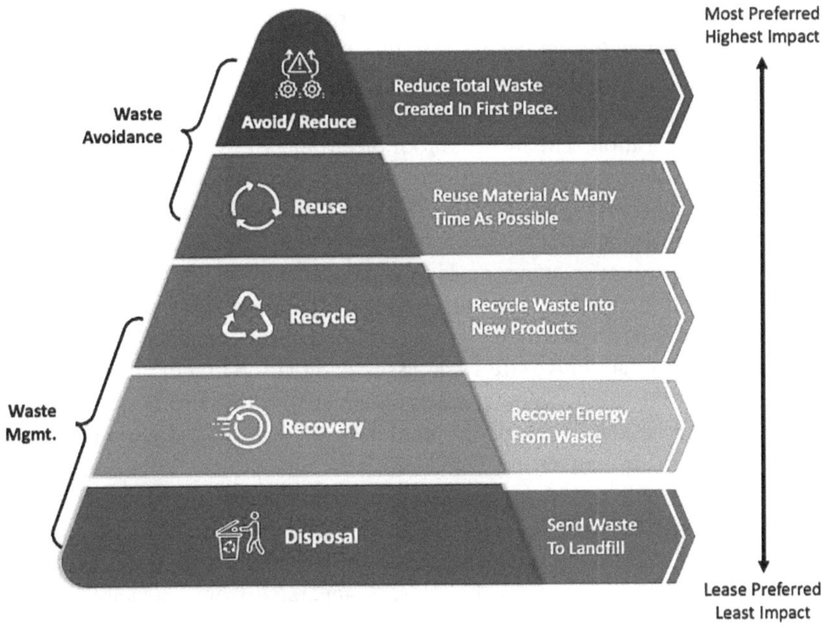

FIGURE 3.3 The hierarchy in waste management, defined by the Waste Framework Directive.

the implementation of efficient strategies for the effective management of technological waste necessitates the adoption of a multi-pronged methodical strategy that adheres to the waste management hierarchy. This hierarchy places significant emphasis on a sequential progression of actions within the realm of waste management, with a primary focus on the minimisation of waste generation, the optimisation of resource use, and the mitigation of environmental damage. The hierarchy consists of five primary stages: Reduce, Reuse, Recycle, Recover, and Dispose (**Figure 3.3**) (Gharfalkar et al. 2015). This section provides a more comprehensive analysis of each step of the hierarchy, emphasising the scientific foundations and practical implementations within various technological waste classifications.

3.3.1 Reduce

At its essence, waste reduction primarily entails the prevention of waste formation at its origin. The primary objective of reduction methods within the domain of metal extraction and mining waste pertains to the enhancement of ore processing methodologies. Sophisticated methods, such as sensor-based ore sorting, employ X-ray transmission and other sensing processes to discern precious minerals and separate them from unproductive rock prior to subsequent processing. An example of such an effective technology in the mining industry is the X-ray transmission-based ore sorting system developed by TOMRA Sorting Solutions (Robben et al. 2020). This technology has been shown to be successful in reducing waste rock during mining operations by accurately recognising precious ore particles. This strategy effectively reduces the influx of non-essential materials into subsequent processes, resulting in a decrease in energy consumption and chemical use (Modise et al. 2022). Furthermore, the utilisation of sophisticated modelling and simulation methodologies can contribute to the development of extraction processes that are characterised by enhanced efficiency and reduced waste generation. The mitigation of electronic waste entails the use of eco-design concepts throughout the process of product development (Gurauskienė and Varžinskas 2006). This involves the process of carefully choosing materials that have a reduced environmental effect, developing goods that can be easily disassembled, and incorporating modular components. The concept of design for disassembly is the deliberate arrangement of components in a manner that enables their removal with the least harm, hence aiding the process of refurbishment and subsequent reuse. The application of quantitative life cycle assessment (LCA) methodologies enables the evaluation of the environmental effect associated with various design options, hence facilitating the identification of opportunities for waste reduction (Hong et al. 2015). Collaborative initiatives like the Open Cell Phone project focus on developing open-source, repairable cell phone designs that emphasise longevity and repairability over disposability. The effective management of plastic waste necessitates a thorough comprehension of the principles and intricacies of polymer chemistry. The investigation of biodegradable and compostable plastics generated from renewable resources is being conducted by researchers with the aim of providing substitutes for conventional plastics. Furthermore, the use of inventive package designs that effectively consume fewer resources while maintaining the quality of the product, in conjunction with the optimisation of distribution networks, has the potential to mitigate waste generation significantly. For example, corporations are now investigating the utilisation of thinner plastic materials while ensuring the preservation of package integrity. The "Plant Bottle" programme of Coca-Cola integrates plant-derived components into PET plastic bottles, therefore diminishing their dependence on polymers derived from fossil fuels (Shogren et al. 2019).

3.3.2 Reuse

The practice of reusing materials and goods may effectively prolong their lives and mitigate the need for new resources. Within the domain of electronic waste, the practice of repairing and reconditioning electronic gadgets serves as a viable means to avert the untimely disposal of such products. Certified technicians are responsible for the diagnosis and repair of problems, the replacement of damaged components, and the verification of the device's performance. The use of standardised testing techniques, as shown by the Mobile Phone Reuse Standard (MPRS), guarantees the adherence of refurbished devices to established quality benchmarks (Dubey 2021). For example, prominent corporations such as Apple and Dell have implemented trade-in initiatives that incentivise users to return their outdated gadgets for refurbishing. This approach effectively mitigates the necessity for additional manufacturing procedures and contributes to resource conservation. Furthermore, the use of reverse logistics systems enables the efficient gathering, conveyance, and redistribution of pre-owned electronic devices for the purpose of reuse. The non-profit organisation "Computers for Schools" in Colombia collects, refurbishes, and donates used computers to schools, extending the life of these devices and bridging the digital divide (Streicher-Porte et al. 2009). The practice of reutilising mining waste entails the identification of alternative uses for materials that would otherwise be disposed away. An example of repurposing mine tailings is their utilisation in the construction of infrastructure such as roads and embankments (Amrani et al. 2019). Geotechnical evaluations and procedures for characterising materials are vital in ascertaining the appropriateness of these materials for reutilisation. The utilisation of geopolymerisation, a technique aimed at transforming waste materials into cementitious substances, is an alternative strategy for the repurposing of mining waste (Segui et al. 2023). The utilisation of mine tailings, residual materials resulting from the extraction and processing of ore, can be employed for several purposes. Boliden, a company operating in Sweden, employs the utilisation of mine tailings for the purpose of road construction, hence mitigating the need for traditional construction materials (Thorén Hedin and Ranängen 2017). This strategy not only involves the repurposing of waste materials but also serves to alleviate the environmental consequences associated with conventional road construction methods. The process of reusing plastic materials frequently entails procedures such as cleaning, disinfecting, and remanufacturing. Nevertheless, the multitude of plastic varieties and their distinct characteristics provide inherent difficulties. The utilisation of sorting technologies, such as near-infrared spectroscopy and automated sorting systems, plays a crucial role in the identification and segregation of diverse plastic materials. The remanufacturing procedures encompass several techniques, such as extrusion, injection moulding, and other methodologies, to convert gathered plastic waste into fresh goods while upholding standards of quality and safety. The "Precious Plastic" initiative offers freely accessible technical designs for compact plastic recycling machinery, empowering nearby communities to convert discarded plastic materials into novel commodities, including home goods and artistic installations (Spekkink, Rödl, and Charter 2022).

3.3.3 Recycle

Recycling is a process that entails transforming waste materials into raw materials or products that possess renewed worth. The appropriate management of plastic waste, which is a significant environmental problem, may be achieved through the practice of recycling. The process of mechanical recycling involves the grinding of plastic waste into granules, which may subsequently be utilised in the production of fresh plastic goods (Ragaert, Delva, and Van Geem 2017). Chemical recycling is a process that involves the decomposition of plastics into monomers or other useful compounds using techniques such as pyrolysis or depolymerisation. Feedstock recycling is a method that transforms plastic waste into intermediate products that may be utilised in various chemical processes. The task at hand involves the development of methodologies capable of effectively managing the wide array of plastic materials found within waste streams while simultaneously maintaining the integrity and quality of the resulting recycled materials. As an illustrative example, it is worth noting that polyethylene terephthalate (PET) plastic, which is frequently seen in beverage bottles, may undergo recycling processes to yield new bottles or alternatively be converted into polyester fibres suitable for textile applications. Coca-Cola, among other corporations, has made a commitment to using a substantial proportion of recycled polyethylene terephthalate (PET) in its packaging materials, therefore mitigating its dependence on newly produced plastic (Ren et al. 2015). The New Plastics Economy effort, spearheaded by the Ellen MacArthur Foundation, aims to advance the use of recyclable plastics and the establishment of closed-loop recycling systems in order to mitigate the generation of plastic waste (Barrowclough and Birkbeck 2022). Metallurgical recycling processes are of utmost importance in the retrieval of precious metals from waste streams. Hydrometallurgical procedures, such as leaching and solvent extraction, are employed to recover metals from waste sources and then isolate them from contaminants (Krishnan et al. 2021). Subsequently, precipitation and electrowinning techniques are utilised to extract metals in a refined state. Efficient metal recovery necessitates meticulous regulation of factors such as pH, temperature, and reagent concentrations throughout these procedures. One such instance is the utilisation of hydrometallurgical techniques to extract copper from mine tailings, generating lucrative copper concentrates and mitigating the ecological consequences associated with traditional mining practices (Falagán, Grail, and Johnson 2017). The recycling of electronic waste involves a range of mechanical and chemical procedures. Mechanical procedures encompass several operations, such as size reduction, shredding, and separation, which are executed through the utilisation of techniques such as air classification and gravity separation. Chemical techniques, such as leaching and smelting, are employed to extract precious metals selectively. Pyrometallurgical techniques, such as high-temperature smelting, are employed for the purpose of extracting metals from circuit boards (Ma 2019). Nevertheless, the intricate composition of electronic waste necessitates a thorough analysis of its materials and the optimisation of processes in order to accomplish efficient recycling. The Urban Mining Company, based in the United States, employs novel methodologies to extract valuable metals from electronic waste, therefore mitigating the necessity for resource-intensive mining and refining processes.

3.3.4 Recover

The process of energy recovery from waste entails the extraction of useful energy, commonly achieved through methods such as incineration or anaerobic digestion. The utilisation of industrial chemical waste, namely organic waste, has the potential to be employed as a source of energy generation. The utilisation of anaerobic digestion in wastewater treatment facilities has been observed through case studies, demonstrating its ability to produce biogas from organic waste constituents. This biogas may then be harnessed to provide both power and heat. It is common for industrial waste streams to contain chemically valuable substances that may be extracted and used for subsequent purposes. Membrane separation methods, such as nanofiltration and reverse osmosis, are utilised for the purpose of selectively segregating valuable components from waste effluents. The chemicals that have been recovered have the potential to be reused in various industrial processes, hence decreasing the need for new resources and mitigating the production of waste. The process of extracting energy from nuclear waste encompasses the utilisation of sophisticated technology such as accelerator-driven systems (ADS) and fast reactors. ADS utilise a particle accelerator to initiate nuclear events within a subcritical reactor, resulting in the transformation of long-lived radioactive isotopes into isotopes that have shorter half-lives or are stable (Gokhale, Deokattey, and Kumar 2006). Fast reactors, which are specifically engineered to function using fast neutrons, have the capability to utilise nuclear waste as a source of fuel effectively. By harnessing the energy from isotopes that would otherwise contribute to prolonged radioactivity, these reactors offer a viable solution for the management of nuclear waste (Kurniawan et al. 2022). Nevertheless, the implementation of these methodologies necessitates meticulous engineering and the implementation of safety protocols to address the inherent complexities associated with radioactive materials effectively.

3.3.5 Dispose

The term "dispose" refers to the act of getting rid of or discarding something. Disposal, the last phase of the hierarchical waste management approach, pertains to the management of waste materials that are not amenable to efficient handling *via* the preceding stages. The proper disposal of nuclear waste is vital due to its intricate and perilous characteristics. The proper management of nuclear waste requires the implementation of rigorous containment and isolation protocols. Deep geological repositories, which are situated in geologically stable rocks, offer a secure setting for the extended containment of radioactive waste. Deep geological repositories, such as the envisaged storage for high-level nuclear waste in Forsmark, Sweden, offer a reliable means of safeguarding radioactive waste by effectively separating it from the biosphere (Chapman and Hooper 2012). Further, the preservation of waste containers over extended periods of time necessitates the implementation of meticulous material selection and the utilisation of designed barrier systems. The examination of natural analogues, such as naturally existing deposits of radioactive materials, aids scientists in comprehending the extended-term dynamics of nuclear waste inside geological formations. The Yucca Mountain repository, situated in the United States, is a planned facility designed to bury high-level nuclear waste below the Earth's surface. Despite encountering political and technical obstacles, the initiative highlights the significance of employing safe disposal techniques for extremely dangerous waste materials (Stuckless and Levich 2016).

The optimum management of technological waste involves a meticulous integration of various aspects of the waste management hierarchy. The successful incorporation of the waste management hierarchy into the processes of metal extraction and mining waste necessitates the collaborative efforts of geologists, engineers, and environmental scientists. Geographic Information Systems (GIS) and geospatial modelling have the ability to assist in the identification of appropriate locations for waste storage facilities. This is achieved by considering various criteria, such as the proximity to mining sites and the potential environmental problems that may arise (Gbanie et al. 2013). Moreover, life cycle assessments have the capability to measure and evaluate the environmental advantages associated with waste reduction, reuse, and recycling techniques in comparison to conventional approaches. Efficient collection, segregation, and processing systems are required for the successful integration of the hierarchy into electronic waste management. The implementation of extended producer responsibility frameworks, wherein producers assume accountability for the disposal of their goods at the end of their lifecycle, serves to promote appropriate disposal practices and streamline recycling processes (Hickle 2014). Technological advancements, such as the use of automated sorting robots and the utilisation of advanced material identification techniques, contribute to the optimisation of recycling facilities by increasing operational efficiency and facilitating the retrieval of valuable materials. The "Green Fence" effort implemented in China serves as a notable illustration of the necessity for appropriate recycling and disposal practices pertaining to electronic waste (Sun 2019). The implementation of more stringent laws regarding the importation of waste products serves the purpose of mitigating environmental degradation while also emphasising the significance of adopting responsible waste management methods on a worldwide scale. The recycling process of rare-earth metals extracted from electronic waste exemplifies the considerable resource potential inherent in recycling practices. The utilisation of these precious metals is crucial for a wide range of advanced technological purposes, and the retrieval of these metals from electronic waste can help mitigate the need for additional mining activities (Kaim, Rintala, and He 2023). The establishment, known as the "E-Waste Recycling Hub", located in Delhi, India, serves as a prime illustration of this necessity for implementing comprehensive waste management practices. The hub utilises shredding, separation, and recovery methodologies to effectively extract valuable elements from electronic waste while also safely handling hazardous components (Dasgupta et al. 2017). The incorporation of the hierarchy into the management of plastic waste necessitates the adoption of a comprehensive and systematic methodology. The fabrication of plastic products is guided by design guidelines that prioritise recyclability. The establishment of uniform labelling for plastic kinds can be facilitated by collaborative endeavours, including plastic producers, recyclers, and legislators. This would contribute to the facilitation of sorting and recycling operations. Furthermore, the utilisation of digital technologies such as blockchain has the potential to augment traceability within the plastic supply chain, therefore guaranteeing the effective oversight and control of plastic waste (Castiglione et al. 2023). Plastic pyrolysis, which is a kind of chemical recycling, facilitates the conversion of plastic waste into economically significant commodities such as fuels and chemicals. The "Chem Cycling" initiative implemented by BASF employs pyrolysis as a means to transform plastic waste into feedstocks that can be utilised by the chemical industry, therefore showcasing the prospect of generating value from discarded materials (Megha et al. 2024).

3.4 SCIENTIFIC ISSUES AND OPPORTUNITIES AROUND TECHNOLOGICAL WASTE MANAGEMENT

Although technological waste offers a plethora of opportunities to realise the idea of "Waste to Treasure", the scientific challenges in their processing and management are both complex and pressing. This section provides an in-depth analysis of the complex scientific obstacles faced by material scientists and their efforts to address various types of technological waste. The integration of diverse expertise from disciplines such as metallurgy, polymer science, electronics, and nuclear engineering assumes great importance in tackling these complex difficulties.

3.4.1 Material Challenges and Prospects in Mining Waste Management

The management of mining waste encounters the formidable task of converting tailings into durable construction materials. Geopolymerisation is a chemical process that entails the interaction between tailings containing high concentrations of aluminosilicates and alkaline activators (He et al. 2022). **Figure 3.4** illustrates the various methods of turning mine tailings into value through geopolymerisation (Xiaolong et al. 2021). Material scientists are engaged in the study of reaction kinetics and the enhancement of activator formulations in order to get certain mechanical properties. Simultaneously, they aim to reduce the environmental effect and guarantee the long-term stability of the materials. The issue of tailings stabilisation extends beyond the scope of geopolymerisation. Material scientists are conducting in-depth investigations into the geomechanical features in order to ensure the ability

FIGURE 3.4 Methods for turning mine tailings into value through geopolymerisation. (Xiaolong et al. 2021.)

of geopolymerised tailings to survive various environmental stresses. Multi-scale modelling methodologies, shown by the implementation of finite element analysis, can facilitate the simulation of tailings behaviour under several loading conditions (Luhar and Luhar 2022). This will enable engineers to effectively devise designs for structures that possess both stability and long-term integrity. Researchers are utilising numerical simulations to examine the distribution of stress within geopolymerised matrices in the setting of copper tailings. This methodology can allow for the enhancement of tailings mixture proportions to provide greater strength and stability while simultaneously mitigating the potential for structural collapse. Material scientists are actively engaged in the exploration of nanotechnology as a means of stabilising tailings. The mechanical properties of geopolymerised tailings are enhanced by the implementation of nanostructured additives by researchers. The manipulation of nanoparticle size and distribution can have a significant impact on the microstructure, leading to enhanced strength and decreased permeability, hence facilitating sustainable waste disposal and recovery (Chougan et al. 2020). The mechanical strength of geopolymerised tailings can be enhanced with the use of nano-silica particles.

The extraction of critical metals from intricate mining waste necessitates the utilisation of specialised separation procedures. The recovery of critical metals also necessitates the use of environmentally friendly and sustainable methodologies (Zhuang et al. 2015). Material scientists are actively developing ligands and adsorbents that possess the ability to bind with specific metals selectively. This binding process facilitates the efficient recovery of these metals from waste solutions. The achievement of enhanced sorbent materials possessing superior selectivity, stability, and recyclability necessitates the deliberate manipulation of material characteristics at the molecular scale. Molecular simulations are being employed to elucidate the mechanics of binding and provide valuable insights for the design of ligands. Scholars are investigating the interactions between ligands and metals in order to improve the effectiveness of adsorption and develop environmentally sustainable techniques for recovery. Through the use of density functional theory (DFT) calculations, researchers are able to obtain a comprehensive understanding of the interactions between ligands and metal ions in terms of electron donation. This knowledge will serve to guide the design of ligands, enabling the achievement of optimal selectivity for certain metal ions. Other innovative approaches are necessary for the extraction of rare earth elements from mining waste. Material scientists are researching electrochemical techniques that have the ability to extract precious metals selectively. Through the manipulation of electrode materials and the optimisation of electrochemical parameters, researchers have successfully facilitated the effective recovery process while simultaneously mitigating any adverse environmental consequences. The development of electrode materials coated with ion-selective polymers has been undertaken by researchers in order to enable the selective extraction of rare earth elements. The utilisation of electrochemical impedance spectroscopy will enable the examination of the interactions between electrodes and electrolytes, hence facilitating the development of recovery techniques that are both efficient and selective.

3.4.2 Material and Design Complexities and Prospects in Electronic Waste Management

The complex nature of electronic equipment poses significant issues in the management of electronic waste. Material scientists are engaging in innovation by employing their expertise to build composite materials that effectively integrate waste components. Nevertheless, the need for comprehensive material characterisation and analysis arises due to compatibility challenges among various waste materials and matrices, along with the possibility of contamination. Material scientists and electrical engineers are also engaging in collaborative efforts to advance the field of sustainable electronics. In the realm of electronic components, there is a growing trend towards the substitution of hazardous chemicals with biocompatible and ecologically benign alternatives. Academic researchers investigate the interactions of materials under diverse settings in order to ascertain the dependability, thermal stability, and electrical efficacy of sustainable electronic gadgets. Material scientists are working on alternate substrates that have a less environmental footprint, examining their thermal, electrical, and mechanical characteristics. Simulations are utilised to examine the behaviour of materials when subjected to various stresses, with the aim of ensuring that sustainable materials are capable of meeting performance standards. The investigation of innovative substrate materials for circuit boards entails the utilisation of finite element models to analyse the dispersion of heat and the impact of mechanical stress (Sri-Jayantha et al. 2008). Computational models are being employed by material scientists to provide predictions regarding thermal conductivity and coefficient of thermal expansion. These predictions serve the purpose of facilitating the identification and selection of substrate materials that can effectively boost electronic performance and durability (Potyrailo et al. 2011). The utilisation of computational modelling techniques can provide a means to get a comprehensive understanding of the underlying mechanisms of ageing and deterioration of electronic circuits and microcomponents. This knowledge can then be leveraged to develop materials that exhibit improved dependability and prolonged lifespans for various devices.

Material scientists are also actively engaged in addressing the complex task of dissolving and then recovering various materials from electronic waste. The utilisation of sophisticated solvent systems and coordination chemistry shows the potential to enable the targeted dissolving of different components. The utilisation of computational modelling here, too, is playing a crucial role in the development of solvent design, enabling the optimisation of recovery procedures for valuable metals and functional compounds. The investigation of ionic liquids with adjustable solvation characteristics is being conducted in order to explore their potential for selectively dissolving metals. Molecular dynamics simulations are being employed to forecast the solvation behaviour of metal ions inside various ionic liquids, hence offering valuable insights into the underlying principles of dissolution (Rode et al. 2005). These customised solvents provide effective separation and retrieval of materials.

The utilisation of lead-free solder alloys with equivalent performance is imperative in meeting the sustainable electronics demand. Material scientists are working on alloy compositions that aim to provide both dependability and mechanical strength while avoiding the harmful consequences associated with lead (Ma and Suhling

2009). The development of lead-free solder materials that are suited for waste reduction and reusability can be guided by thermodynamic modelling and experimental validation. The eutectic solder alloy composed of tin, silver, and copper has emerged as a viable lead-free option (Abtew and Selvaduray 2000). Researchers are employing CALPHAD (Calculation of Phase Diagrams) modelling to provide predictions regarding the solidification behaviour and mechanical characteristics (Ågren 1996). This methodology will guarantee that lead-free solder materials conform to performance criteria while also mitigating the production of hazardous waste derived from electronic devices.

3.4.3 Challenges and Opportunities in Plastic Waste Management

Polymer scientists are actively addressing the inherent compatibility issues associated with the amalgamation of diverse plastic waste streams with the aim of generating novel materials. The utilisation of melt blending techniques and the invention of compatibilisers facilitate the efficient mixing of materials while preserving the required properties (Yin et al. 2015). The comprehension of intermolecular interactions and phase behaviour plays a crucial role in the ability to forecast material behaviour and enhance blend compositions. Molecular dynamics simulations are being utilised by polymer scientists in order to investigate the molecular-level behaviour of polymer blends. The comprehension of chain interactions, phase separation, and domain morphology plays a crucial role in informing the choice of suitable polymers for efficient waste blending (Prathab, Subramanian, and Aminabhavi 2007). Computational insights will play a crucial role in the optimisation of blend compositions and processing conditions. The mixing behaviour of various polymer waste kinds, such as polyethylene terephthalate (PET) and high-density polyethylene (HDPE), is being investigated through the application of molecular dynamics simulations by researchers (Mashaan, Chegenizadeh, and Nikraz 2022). Through the examination of intermolecular forces and segmental mobility, researchers are able to determine the most favourable blend ratios that lead to enhanced mechanical characteristics and less phase separation.

The process of biodegradation of plastic waste is dependent on enzyme reactions that facilitate the breakdown of polymers (Brodhagen et al. 2015). Computational modelling can also be employed by material scientists to examine the interactions between enzymes and substrates, with the aim of elucidating the processes of degradation. The understanding of this information can drive the development of enzymes that exhibit improved activity and selectivity, hence propelling the progress of biodegradation technologies. Enzymes are being engineered by material scientists to achieve improved levels of activity and selectivity. Polymer scientists are also engaged in research and development activities pertaining to the formulation of biodegradable polymer blends, with the aim of facilitating sustainable management of plastic waste. The utilisation of biodegradable polymers in conjunction with synthetic plastics enables the regulation of breakdown processes (Siracusa et al. 2008). Material scientists are also investigating the rates of mix deterioration, mechanical integrity, and environmental impact in order to develop customised materials that may effectively mitigate plastic pollution. The investigation of controlled degradation is being conducted on blends of polylactic acid (PLA) with synthetic polymers. Thermal analysis techniques are utilised by researchers to observe the degradation process, therefore providing insights into the impact of blend composition and ambient circumstances on the kinetics of degradation (Reddy, Nama, and Yang 2008). This knowledge can assist in the development of mixes with specific degrading characteristics.

3.4.4 Challenges and Opportunities With Nuclear Waste Materials and Long-Term Containment

Material scientists are also responsible for the design of engineered barriers used in nuclear waste disposal. The customisation of immobilisation matrices, such as glass and ceramics, is necessary in order to handle waste isotopes effectively. Thermodynamic modelling is being utilised to forecast the dissolution of glass and the stability of ceramics in repository environments, assuring the long-term integrity of barriers for containment purposes. Besides, the construction of nuclear waste storage buildings necessitates the use of materials that possess resistance to radiation (Davidovits 1994). Alloys capable of withstanding deterioration caused by radiation are being developed. The utilisation of irradiation tests and microstructural characterisation techniques can provide valuable insights into the behaviour of alloys when subjected to harsh circumstances (Knaster, Moeslang, and Muroga 2016). This knowledge is crucial in the development of durable structural materials that can effectively retain waste. Neutron irradiation can be employed as a means of evaluating the radiation-induced hardening and embrittlement of austenitic stainless steels (Busby, Hash, and Was 2005). Transmission electron microscopy can also be utilised by material scientists for the purpose of visualising defect clusters that are generated by irradiation. These observations will provide guidance for the modification of alloys in order to limit the effects of radiation, thus assuring the long-term durability of structures in repositories.

Material scientists are also doing research and working on the development of alloys that exhibit exceptional resistance to radiation and possess robust mechanical properties (Yvon and Carré 2009). The evaluation of material performance over extended periods necessitates the use of *in-situ* testing to examine radiation-induced alterations in material characteristics, such as embrittlement. Material scientists are actively engaged in addressing the durability difficulties associated with repository materials. Reactive transport simulations are being utilised to explore the enduring interactions between constructed barriers and waste isotopes. The models employed in this study aim to forecast the release of radionuclides and the potential transformation of barriers, with the ultimate objective of maintaining the long-term integrity and functionality of materials over extended periods of thousands of years. Simulation models that utilise reactive transport equations are utilised to provide predictions about the behaviour of glass waste forms in repository settings. The analysis of radioactive diffusion through the glass matrix over longer periods of time is conducted by researchers (Gin et al. 2017). This analysis facilitates the determination of appropriate glass compositions that may effectively isolate waste for long durations. In addition to these, self-healing materials for nuclear waste deposit barriers are also a subject of investigation by material scientists. Barrier integrity can be restored by the activation of autonomous repair mechanisms in response to damage caused by radiation. Researchers are investigating self-healing agents and the material's reaction to radiation in order to improve the long-term confinement of radioactive waste. The integration of microcapsules containing healing chemicals into barrier materials facilitates self-repair mechanisms. The microcapsules are ruptured as a result of radiation-induced damage, leading to the release of healing chemicals that effectively close cracks and preserve the integrity of the barrier (White et al. 2011). The novel methodology exhibits potential for the sustainable containment of nuclear waste.

3.5 GLOBAL REGULATIONS AND LEGISLATIONS AFFECTING TECHNOLOGICAL WASTE MANAGEMENT

The intricate relationship among environmental preservation, the retrieval of resources, and adherence to industrial regulations highlights the intricate nature of waste management in contemporary society. Within the complex framework of technological waste management, the inclusion of worldwide regulations and legislation plays a pivotal role in interconnecting the many components of conscientious custodianship and ecological durability. The influence of global rules and legislation on waste management methods is significant, affecting several areas. This section explores several regulatory frameworks from across the world that influence the governance of various types of technological waste.

3.5.1 Need for Regulations: Challenges in Waste Management

Prior to learning about various important regulations surrounding technological waste management, it is imperative to understand the necessity for these regulations. This primarily stems from the adverse effects of leniency or the absence of regulatory frameworks to tackle waste management on the environment and humans. Artisanal and small-scale mining frequently takes place in areas characterised by inadequate regulatory frameworks. One of the adverse consequences is the inadequate disposal of waste generated during the process of metal extraction, resulting in the pollution of soil and water resources (Lu et al. 2015). The absence of proper supervision amplifies the deterioration of the environment and the associated health hazards. In certain nations, the practice of artisanal gold mining leads to the unregulated discharge of tailings polluted with mercury into river systems. The lack of efficient regulatory enforcement results in the contribution of mercury pollution and the occurrence of harmful health consequences on nearby populations (Esdaile and Chalker 2018). Certain geographical areas also allow for the deposition of metal extraction tailings in natural aquatic environments, resulting in the contamination of aquatic ecosystems. The consequences encompass sedimentation, degradation of water quality, and the loss of habitats. The historical practice of depositing mine tailings in the water bodies of Butte, Montana, USA, resulted in the introduction of heavy metals into the local rivers and lakes, causing pollution (Nimick and Moore 1991). The aquatic environment saw a decline in biodiversity, thereby illustrating the ecological ramifications associated with insufficient waste management rules.

The presence of lenient restrictions in certain locations facilitates the exportation of electronic waste from industrialised countries to underdeveloped nations, hence engendering environmental and health risks. Negative consequences arise from the practice of open burning of electronic waste, which results in the emission of hazardous chemicals and poses a threat to nearby populations. The Agbogbloshie electronic waste dump located in Ghana serves as a symbol of the widespread practice of e-waste disposal. Unregulated recycling processes have been shown to emit toxic compounds into the atmosphere, land, and water, posing a significant risk to the well-being of both employees involved in the recycling process and anyone residing in close proximity to these operations (Yang et al. 2020). Insufficient regulatory measures pertaining to the exportation of plastic waste lead to the buildup of waste inside certain nations. This phenomenon is a significant factor in the occurrence of marine pollution and the deterioration of the ecosystem.

The lack of adequate accountability is a hindrance to the successful implementation of waste management strategies. The prohibition imposed by China on the importation of plastic waste in 2018 has had a significant impact on the worldwide movement of waste materials (Yoshida 2022). Countries that export waste had difficulties in identifying alternate destinations, thereby emphasising the necessity for enhanced rules and more investment in domestic recycling infrastructure. Certain governments have identified loopholes within their waste export laws that enable waste to be reclassified as "commodities" eligible for trade. One of the adverse consequences is the practice of disguising waste dumping as recycling activities in poor nations. This phenomenon contributes to the escalation of environmental contamination and poses potential health hazards.

3.5.2 Positive Regulations: Fostering Sustainable Waste Management

- Mercury is frequently employed in the extraction of gold from ore within the artisanal gold mining sector in underdeveloped nations. The **Minamata Convention** (Kessler 2013) promotes the utilisation of more environmentally friendly techniques for gold extraction, such as gravity concentration and cyanide leaching, in order to mitigate mercury emissions and protect human health.
- The **Extractive Industries Transparency Initiative (EITI)** (Corrigan 2014) promotes transparency within the extractive industry, with a specific focus on ensuring that the money generated from metal extraction is effectively utilised to benefit the local populations. The initiative aims to foster appropriate waste management techniques and encourage community involvement. The EITI in Burkina Faso promotes the practice of mining corporations revealing their financial transactions and payments made to the government. The promotion of openness in waste management and environmental protection cultivates

public confidence and enables communities to effectively monitor and hold firms and authorities responsible for their obligations in these areas.

- The **Basel Convention** (Wath et al. 2010) is a regulatory framework that governs the cross-border transportation of hazardous waste, with the primary objective of mitigating the occurrence of inappropriate waste disposal practices. The convention serves as a platform for fostering international collaboration and facilitating the sharing of knowledge in order to further the development of sustainable waste management solutions. The involvement of the Basel Convention resulted in the repatriation of electronic waste from industrialised nations back to their respective places of origin. This transition guarantees appropriate handling and disposal methods, hence mitigating the environmental and health consequences experienced by developing nations.
- The **Waste Framework Directive** defines a hierarchical framework for the management of waste, with a particular emphasis on prioritising prevention, reuse, recycling, and recovery (Gharfalkar et al. 2015). The use of this comprehensive strategy promotes the reduction of waste and the conservation of resources. The development of waste management plans by member states serves to promote sustainable practices and support innovation in this field. The adoption of expanded producer responsibility under the Directive in Germany has resulted in the promotion of electronic waste recycling. Manufacturers bear the responsibility of undertaking the collection and recycling of their goods, alleviating the strain on municipal waste management systems and advancing the ideas of a circular economy (Cahill, Grimes, and Wilson 2010).
- The **Resource Conservation and Recovery Act (RCRA)** (Harthill 2020) in the United States is a regulatory framework that aims to facilitate the secure handling of hazardous waste, with a particular focus on minimising waste generation at its source, promoting recycling practices, and ensuring appropriate disposal methods. The legislation promotes the adoption of more environmentally friendly manufacturing procedures within industries, resulting in a reduction in the output of waste. Pharmaceutical manufacturers have implemented environmentally conscious measures in accordance with the RCRA regulations for hazardous waste generators. These measures include the adoption of sustainable practices aimed at reducing waste generation during production processes, as well as the implementation of solvent recycling systems.
- The **Containers and Packaging Recycling Act** in Japan aims to address the issue of packaging waste by the implementation of compulsory recycling rates that firms are required to meet. The legislation promotes the use of eco-design techniques and incentivises firms to reduce the use of packaging materials. The effective implementation of obligatory package recycling in Japan resulted in the construction of a robust recycling infrastructure (Usui 2008). This exemplifies the manner in which regulatory frameworks may facilitate the implementation of effective waste management strategies.

- Regulatory measures such as the **Dodd-Frank Wall Street Reform** and **Consumer Protection Act** aim to tackle the utilisation of conflict minerals within the electronics industry (Acharya et al. 2011). These restrictions serve to encourage ethical sourcing practices and mitigate the financing of armed conflict in places abundant in minerals. The advocacy for minerals free from violence in the Democratic Republic of Congo underscores the beneficial outcomes of regulatory measures (Diemel and Hilhorst 2019). Electronics corporations actively participate in supply chain due diligence, therefore demonstrating their commitment to promoting conflict-free mineral commerce and fostering regional stability.
- **Extended Producer obligation (EPR)** (Cahill, Grimes, and Wilson 2010) is a policy approach that imposes the obligation on producers to manage and address the environmental impacts of their goods across the whole lifespan, encompassing manufacturing, usage, and disposal stages. Positive results encompass the provision of incentives for the purpose of encouraging the development of environmentally friendly product designs, as well as the facilitation of the widespread use of recycling technology. The implementation of the Electronic Product Recycling (EPR) programme in South Korea prompted manufacturers to undertake product redesign efforts aimed at facilitating disassembly and recycling processes. Not only did this result in a reduction in electronic waste, but it also fostered the emergence of potential for the development of innovative recycling technologies (Hyunmyung and Yong-Chul 2006).

Apart from the above discussed regulations, There are numerous numbers of other international organisations and initiatives shown in **Table 3.1** which have been a step forward for suitable monitoring and recycling of electronic waste (Sthiannopkao and Wong 2013; Ilankoon et al. 2018).

3.5.3 Finding Balance Between Environmental Protection and Innovation

The discussion suggests that there is a need to strike a balance between innovation in technological waste management and environmental preservation. The long-term and sustainable solution to this is the development of more and more green technologies to manage and beneficiate technological waste. However, in the present scenario, several other considerations can be effective in finding balance around this dilemma. The requirement to get the free, prior, and informed consent of indigenous populations in relation to mining endeavours serves as a means to strike a balance between safeguarding the environment and promoting economic progress. Positive benefits can be observed through the integration of traditional knowledge into waste management strategies, as well as the reduction of adverse effects on indigenous areas. First Nations communities in Canada have expressed apprehensions over the discharge of mining waste into water basins in close proximity to their territories. Achieving a harmonious equilibrium between economic progress and the protection of indigenous rights requires the implementation of cooperative strategies

TABLE 3.1
Initiatives and legislation addressing electronic waste management

Initiatives	Key Features
Bamako Convention	Aimed to restrain the import of e-waste more stringently than the Basel Convention. Applied in African Union nations from 1998.
EU WEEE Directive	In 2007, all the EU members adopted the system with the initiation of a takeback approach for 10 groups of electrical equipment.
Restriction of Hazardous Substances Directive (RoSH)	Enforced along with EU WEEE, particularly aimed to restrain the use of hazardous substances. It is also validated by various nations, including China and India.
Solving the E-waste Problem (StEP)	Initiated in 2007 by UN agencies to promote the reusability of recycled components to limit waste generation.
3Rs (Reduce, Reuse, Recycle)	Initiated by Japan. Work to prevent e-waste generation. Allows exportation to other countries for remanufacture and recycling. Conflicting the goal of the Basel Convention treaty.
US State laws and the Responsible Electronic Recycling Act (HR2284)	25 states of the US, imposed with the law. HR2284 is anticipated to control e-waste exportation. This law enforces the assembly and reprocessing of e-waste via stipulating deposits from consumers.
US NGOs–Basel Action Network (BAN), Silicon Valley Toxic Coalition (SVTC), Electronics Takeback Coalition (ETBC)	These three acts work to promote the "Basel Ban" amendment for restrictive transborder exportation. Enhance general e-waste assortment and reusing programmes.
National Strategy for Electronics Stewardship (NSES), US	Focus on limiting the use of harmful substances. Improve the handling and management strategies of e-waste in the US or reduce their harmful impact in other nations.
International Environmental Technology Centre (IETC) 7– UNEP	Strengthen utilisation of environmentally suitable technologies in developing nations on waste management.
Global e-Sustainability Initiative (GeSI)	Focus on engaging Information and Communication Technology (ICT) companies, industries and organisations for e-waste management.

Sources: Sthiannopkao and Wong 2013; Ilankoon et al. 2018.

that take into account both traditional knowledge and contemporary waste management methodologies. The Directive on the Restriction of Hazardous Substances (RoHS) imposes limitations on the utilisation of dangerous substances in electronic goods (Cheng, Li, and Huang 2023). The mitigation of environmental contamination poses hurdles in terms of compliance, mostly stemming from the intricate nature of supply chains and the necessity for material substitutes. The adherence to the RoHS directive has resulted in the elimination of lead from electronic devices, therefore mitigating potential health hazards. The primary objective of the European Union's REACH (Registration, Evaluation, Authorisation, and Restriction of Chemicals) initiative is to guarantee the secure utilisation of chemicals, all the while fostering advancements in the field. The issues associated with REACH compliance stem from the need for data and the financial obligations it imposes, which may pose obstacles for small-scale innovators, notwithstanding the emphasis on promoting openness and hazard assessment. The REACH legislation facilitates the replacement of hazardous chemicals with less harmful alternatives (Sarkar et al. 2021). Nevertheless, small and medium firms encounter difficulties in fulfilling the financial obligations associated with compliance and the demands of data sharing, which present obstacles to the advancement of innovation.

3.6 SUMMARY

This chapter has explored a terrain where the complex interplay of science, industry, legislation, and environmental awareness intersect—a terrain that delineates the crucial significance of waste management in our pursuit of a sustainable future. The discussion started with an exploration of the technological waste management hierarchy, which consists of the following tiers: Reduce, Reuse, Recycle, Recover, and Disposal. The discussion delved into each tier, examining the tactics employed, as well as the scientific foundations and practical implementations that underpin them. The scientific discourse has shown the noteworthy accomplishments of material scientists, engineers, and inventors who have utilised their expertise to convert waste materials into useful resources. The domain of scientific inquiry is not the only confinement of the intricate fabric of technological waste management. The influence of positive and negative directions in waste management techniques may be observed in the realm of worldwide regulations and legislation, which encompasses a wide range of policies and guidelines. By examining real-world case studies, the chapter observed the significant influence that regulatory frameworks may have, highlighting the potential for positive change, the negative outcomes that can arise from insufficient control, and the practical potential offered by the concepts of a circular economy. This chapter is a fundamental component in the development of sustainable advancement, wherein the harmonious orchestration of waste management reverberates with cohesion rather than dissonance. This statement underscores the inherent compatibility between our technological advancements and our dedication to preserving the environment. Through the adoption and integration of comprehensive waste management practices, we provide the groundwork for a symbiotic relationship between technological advancements, the preservation of resources, and the maintenance of ecological balance.

Ultimately, the management of technological waste extends beyond the boundaries of scientific inquiry, regulatory frameworks, and industrial practices. This highlights the remarkable ability of the human spirit to persist and endure. It emphasises the spirit's capacity to imagine a future in which growth and the protection of the environment are harmoniously intertwined. Furthermore, it underscores the role of innovation as a driving force behind the responsible management of natural resources.

REFERENCES

Abtew, Mulugeta, and Guna Selvaduray. 2000. Lead-free Solders in Microelectronics. *Materials Science and Engineering: R: Reports* 27 (5):95–141.

Acharya, Viral V., Thomas Cooley, Matthew Richardson, Richard Sylla, and Ingo Walter. 2011. The Dodd-Frank wall street reform and consumer protection act: Accomplishments and limitations*. *Journal of Applied Corporate Finance* 23 (1):43–56.

Adam, Issahaku, Tony R. Walker, Joana Carlos Bezerra, and Andrea Clayton. 2020. Policies to reduce single-use plastic marine pollution in West Africa. *Marine Policy* 116:103928.

Ågren, John. 1996. Calculation of phase diagrams: Calphad. *Current Opinion in Solid State and Materials Science* 1 (3):355–360.

Ambaye, Teklit Gebregiorgis, Mentore Vaccari, Francine Duarte Castro, Shiv Prasad, and Sami Rtimi. 2020. Emerging technologies for the recovery of rare earth elements (REEs) from the end-of-life electronic wastes: A review on progress, challenges, and perspectives. *Environmental Science and Pollution Research* 27 (29):36052–36074.

Amrani, Mustapha, Yassine Taha, Azzouz Kchikach, Mostafa Benzaazoua, and Rachid Hakkou. 2019. Valorization of phosphate mine waste rocks as materials for road construction. *Minerals* 9(4):237.

Bahadur, Singh Nakshatra, and Agarwal Sonal. 2016. Nanocomposites: An overview. *Emerging Materials Research* 5 (1):5–43.

Barrowclough, Diana, and Carolyn D. Birkbeck. 2022. Transforming the global plastics economy: The role of economic policies in the global Governance of plastic pollution. *Social Sciences* 11 (1):26.

Brodhagen, Marion, Mark Peyron, Carol Miles, and Debra Ann Inglis. 2015. Biodegradable plastic agricultural mulches and key features of microbial degradation. *Applied Microbiology and Biotechnology* 99 (3):1039–1056.

Busby, Jeremy T., Mark C. Hash, and Gary S. Was. 2005. The relationship between hardness and yield stress in irradiated austenitic and ferritic steels. *Journal of Nuclear Materials* 336 (2):267–278.

Cahill, Rachel, Sue M. Grimes, and David C. Wilson. 2010. Review article: Extended producer responsibility for packaging wastes and WEEE – A comparison of implementation and the role of local authorities across Europe. *Waste Management & Research* 29 (5):455–479.

Castiglione, Aniello, Lucia Cimmino, Mario Di Nardo, and Teresa Murino. 2023. A framework for achieving a circular economy using the blockchain technology in a sustainable waste management system. *Computers & Industrial Engineering* 180:109263.

Chapman, Neil, and Alan Hooper. 2012. The disposal of radioactive wastes underground. *Proceedings of the Geologists' Association* 123 (1):46–63.

Cheng, Jeng-Chieh, Jeen-Fong Li, and Chi-Yo Huang. 2023. Enablers for adopting restriction of hazardous substances directives by electronic manufacturing service providers. *Sustainability* 15 (16):12341.

Chougan, Mehdi, Seyed Hamidreza Ghaffar, Mohammad Jahanzat, Abdulrahman Albar, Nahzatullah Mujaddedi, and Rafiq Swash. 2020. The influence of nano-additives in strengthening mechanical performance of 3D printed multi-binder geopolymer composites. *Construction and Building Materials* 250:118928.

Corrigan, Caitlin C. 2014. Breaking the resource curse: Transparency in the natural resource sector and the extractive industries transparency initiative. *Resources Policy* 40:17–30.

Dasgupta, Dipsikha, Anupam Debsarkar, Tumpa Hazra, B. K. Bala, Amitava Gangopadhyay, and Debasish Chatterjee. 2017. Scenario of future e-waste generation and recycle-reuse-landfill-based disposal pattern in India: A system dynamics approach. *Environment, Development and Sustainability* 19 (4):1473–1487.

Davidovits, Joseph. 1994. Recent progresses in concretes for nuclear waste and uranium waste containment. *Concrete International* 16 (12):53–58.

Debnath, Biswajit, Abhijit Das, Potluri Anil Chowdary, and Siddhartha Bhattacharyya. 2023. *Development in E-Waste Management: Sustainability and Circular Economy Aspects*: CRC Press.

Demirbas, Ayhan. 2011. Waste management, waste resource facilities and waste conversion processes. *Energy Conversion and Management* 52 (2):1280–1287.

Diemel, J. A., and D. J. M. Hilhorst. 2019. Unintended consequences or ambivalent policy objectives? Conflict minerals and mining reform in the Democratic Republic of Congo. *Development Policy Review* 37 (4):453–469.

Dubey, Monika. 2021. A review of harmful aspects of mobile on health and environment. *Indian Journal of Public Health Research Development* 12 (1), 538-544.

Esdaile, Louisa J., and Justin M. Chalker. 2018. The mercury problem in artisanal and small-scale gold mining. *Chemistry – A European Journal* 24 (27):6905–6916.

Falagán, Carmen, Barry M. Grail, and D. Barrie Johnson. 2017. New approaches for extracting and recovering metals from mine tailings. *Minerals Engineering* 106:71–78.

Gbanie, Solomon Peter, Paul Bobby Tengbe, Jinnah Samuel Momoh, James Medo, and Victor Tamba Simbay Kabba. 2013. Modelling landfill location using Geographic Information Systems (GIS) and Multi-Criteria Decision Analysis (MCDA): Case study Bo, Southern Sierra Leone. *Applied Geography* 36:3–12.

Gharfalkar, Mangesh, Richard Court, Callum Campbell, Zulfiqur Ali, and Graham Hillier. 2015. Analysis of waste hierarchy in the European waste directive 2008/98/EC. *Waste Management* 39:305–313.

Gin, Stephane, Patrick Jollivet, Magaly Tribet, Sylvain Peuget, and Sophie Schuller. 2017. Radionuclides containment in nuclear glasses: An overview. *Radiochimica Acta*, 105 (11): 927–959.

Gokhale, Pratibha A., Sangeeta Deokattey, and Vijai Kumar. 2006. Accelerator driven systems (ADS) for energy production and waste transmutation: International trends in R&D. *Progress in Nuclear Energy* 48 (2):91–102.

Gurauskienė, Inga, and Visvaldas Varžinskas. 2006. Eco-design methodology for electrical and electronic equipment industry. *Environmental Research, Engineering Management* 37 (3):43–51.

Harthill, Michalann. 2020. *Hazardous Waste Management: In Whose Backyard?*: Routledge.

He, Xiang, Zuhua Yuhua, Shaker Qaidi, et al. 2022. Mine tailings-based geopolymers: A comprehensive review. *Ceramics International* 48 (17):24192–24212.

Hickle, Garth T. 2014. An examination of governance within extended producer responsibility policy regimes in North America. *Resources, Conservation and Recycling* 92:55–65.

Hong, Jinglan, Wenxiao Shi, Yutao Wang, Wei Chen, and Xiangzhi Li. 2015. Life cycle assessment of electronic waste treatment. *Waste Management* 38:357–365.

Hyunmyung, Yoon, and Jang Yong-Chul. 2006. The practice and challenges of electronic waste recycling in Korea with emphasis on Extended Producer Responsibility (EPR). *Paper Read at Proceedings of the 2006 IEEE International Symposium on Electronics and the Environment, 8–11 May 2006, Scottsdale, AZ, USA, 2006,* pp. 326–330.

Ilankoon, I. M. S. K., Yousef Ghorbani, Meng Nan Chong, Gamini Herath, Thandazile Moyo, and Jochen Petersen. 2018. E-waste in the international context – A review of trade flows, regulations, hazards, waste management strategies and technologies for value recovery. *Waste Management* 82:258–275.

Islam, Aminul, Tofayal Ahmed, Md Rabiul Awual, et al. 2020. Advances in sustainable approaches to recover metals from e-waste-A review. *Journal of Cleaner Production* 244:118815.

Jaibee, Shafizan, Abd Khalil Abd Rahim, Fariza Mohamad, et al. 2015. Review on current status of waste electric and electronic product in Malaysia. *Applied Mechanics and Materials* 773–774:898–907.

Kaim, Vishakha, Jukka Rintala, and Chao He. 2023. Selective recovery of rare earth elements from e-waste via ionic liquid extraction: A review. *Separation and Purification Technology* 306:122699.

Kessler, Rebecca. 2013. The Minamata convention on mercury: A first step toward protecting future generations. *Environmental Health Perspectives* 121 (10):A304–A309.

Knaster, J., A. Moeslang, and T. Muroga. 2016. Materials research for fusion. *Nature Physics* 12 (5):424–434.

Koutinas, Apostolis A, Anestis Vlysidis, Daniel Pleissner, et al. 2014. Valorization of industrial waste and by-product streams via fermentation for the production of chemicals and biopolymers. *Chemical Society Reviews* 43 (8):2587–2627.

Krishnan, Santhana, Nor Syahidah Zulkapli, Hesam Kamyab, et al. 2021. Current technologies for recovery of metals from industrial wastes: An overview. *Environmental Technology & Innovation* 22:101525.

Kumar, Sunil, and Anupama Sharma. 2019. Cadmium toxicity: Effects on human reproduction and fertility. Reviews on Environmental Health 34 (4):327–338.

Kurniawan, Tonni Agustiono, Mohd Hafiz Dzarfan Othman, Deepak Singh, et al. 2022. Technological solutions for long-term storage of partially used nuclear waste: A critical review. *Annals of Nuclear Energy* 166:108736.

Lomwongsopon, Passanun, and Cristiano Varrone. 2022. Critical review on the progress of plastic bioupcycling technology as a potential solution for sustainable plastic waste management. *Polymers* 14 (22):4996.

Lu, Yonglong, Shuai Song, Ruoshi Wang, et al. 2015. Impacts of soil and water pollution on food safety and health risks in China. *Environment International* 77:5–15.

Luhar, Ismail, and Salmabanu Luhar. 2022. A comprehensive review on fly ash-based geopolymer. *Journal of Composites Science* 6 (8):219.

Ma, En. 2019. Chapter 11 – Recovery of waste printed circuit boards through pyrometallurgy. In *Electronic Waste Management and Treatment Technology*, edited by M. N. V. Prasad, and M. Vithanage: Butterworth-Heinemann.

Ma, Hongtao, and Jeffrey C. Suhling. 2009. A review of mechanical properties of lead-free solders for electronic packaging. *Journal of Materials Science* 44 (5):1141–1158.

Martins, Natalia P., Sumit Srivastava, Francisco V. Simão, et al. 2021. Exploring the potential for utilization of medium and highly sulfidic mine tailings in construction materials: A review. *Sustainability* 13 (21):12150.

Mashaan, Nuha, Amin Chegenizadeh, and Hamid Nikraz. 2022. A comparison on physical and rheological properties of three different waste plastic-modified bitumen. *Recycling* 7 (2):18.

Megha, Mahajan, Murugesan Kamaraj, Thirumullaivoyal G. Nithya, Shanmugaselvam GokilaLakshmi, Pugazh Santhosh, and Balasubramanian Balavaishnavi. 2024. *Biodegradable Polymers – Research and Applications. Physical Sciences Reviews* 9 (2):949–972.

Mikula, Katarzyna, Dawid Skrzypczak, Grzegorz Izydorczyk, et al. 2021. 3D printing filament as a second life of waste plastics—A review. *Environmental Science and Pollution Research* 28 (10):12321–12333.

Modise, E. G., A. M. Zungeru, B. Mtengi, and A. U. Ude. 2022. Sensor-based ore sorting—A review of current use of electromagnetic spectrum in sorting. *IEEE Access* 10:112307–112326.

Nimick, David A., and Johnnie N. Moore. 1991. Prediction of water-soluble metal concentrations in fluvially deposited tailings sediments, upper Clark Fork Valley, Montana, U.S.A. *Applied Geochemistry* 6 (6):635–646.

Potyrailo, Radislav, Krishna Rajan, Klaus Stoewe, Ichiro Takeuchi, Bret Chisholm, and Hubert Lam. 2011. Combinatorial and high-throughput screening of materials libraries: Review of state of the art. *ACS Combinatorial Science* 13 (6):579–633.

Prathab, B., V. Subramanian, and T. M. Aminabhavi. 2007. Molecular dynamics simulations to investigate polymer–polymer and polymer–metal oxide interactions. *Polymer* 48 (1):409–416.

Ragaert, Kim, Laurens Delva, and Kevin Van Geem. 2017. Mechanical and chemical recycling of solid plastic waste. *Waste Management* 69:24–58.

Reddy, Narendra, Digvijay Nama, and Yiqi Yang. 2008. Polylactic acid/polypropylene polyblend fibers for better resistance to degradation. *Polymer Degradation and Stability* 93 (1):233–241.

Ren, Haiyu, Fangqing Qiao, Yu Shi, et al. 2015. PlantBottle™ Packaging program is continuing its journey to pursue bio-mono-ethylene glycol using agricultural waste. *Journal of Renewable and Sustainable Energy* 7 (4):041510.

Robben, Christopher, Pedro Condori, Angel Pinto, Ronald Machaca, and Anssi Takala. 2020. X-ray-transmission based ore sorting at the San Rafael tin mine. *Minerals Engineering* 145:105870.

Rode, B. M., C. F. Schwenk, T. S. Hofer, and B. R. Randolf. 2005. Coordination and ligand exchange dynamics of solvated metal ions. *Coordination Chemistry Reviews* 249 (24):2993–3006.

Sarkar, Swagata, Juliana Dias Bernardes Gil, James Keeley, and Kees Jansen. 2021. *The Use of Pesticides in Developing Countries and Their Impact on Health and the Right to Food*: European Union.

Segui, Pauline, Amine E. Safhi, Mustapha Amrani, and Mostafa Benzaazoua. 2023. Mining wastes as road construction material: A review. *Minerals* 13 (1):90.

Sethurajan, Manivannan, Eric D. van Hullebusch, Danilo Fontana, et al. 2019. Recent advances on hydrometallurgical recovery of critical and precious elements from end of life electronic wastes – A review. *Critical Reviews in Environmental Science and Technology* 49 (3):212–275.

Shogren, Randal, Delilah Wood, William Orts, and Gregory Glenn. 2019. Plant-based materials and transitioning to a circular economy. *Sustainable Production and Consumption* 19:194–215.

Siddique, Rafat, Jamal Khatib, and Inderpreet Kaur. 2008. Use of recycled plastic in concrete: A review. *Waste Management* 28 (10):1835–1852.

Siracusa, Valentina, Pietro Rocculi, Santina Romani, and Marco Dalla Rosa. 2008. Biodegradable polymers for food packaging: A review. *Trends in Food Science & Technology* 19 (12):634–643.

Spekkink, Wouter, Malte Rödl, and Martin Charter. 2022. Repair Cafés and precious plastic as translocal networks for the circular economy. *Journal of Cleaner Production* 380:135125.

Sri-Jayantha, S. M., G. McVicker, K. Bernstein, and J. U. Knickerbocker. 2008. Thermomechanical modeling of 3D electronic packages. *IBM Journal of Research and Development* 52 (6):623–634.

Sthiannopkao, Suthipong, and Ming Hung Wong. 2013. Handling e-waste in developed and developing countries: Initiatives, practices, and consequences. *Science of the Total Environment* 463–464:1147–1153.

Streicher-Porte, Martin, Christian Marthaler, Heinz Böni, Mathias Schluep, Ángel Camacho, and Lorenz M. Hilty. 2009. One laptop per child, local refurbishment or overseas donations? Sustainability assessment of computer supply scenarios for schools in Colombia. *Journal of Environmental Management* 90 (11):3498–3511.

Stuckless, John S., and Robert A. Levich. 2016. The road to Yucca mountain—Evolution of nuclear waste disposal in the United States. *Environmental & Engineering Geoscience* 22 (1):1–25.

Sun, Meng. 2019. The effect of border controls on waste imports: Evidence from China's Green Fence campaign. *China Economic Review* 54:457–472.

Thorén Hedin, Lisa, and Helena Ranängen. 2017. Community involvement and development in Swedish mining. *The Extractive Industries and Society* 4 (3):630–639.

Tian, Quyang, Piao Xu, Danlian Huang, et al. 2023. The driving force of biomass value-addition: Selective catalytic depolymerization of lignin to high-value chemicals. *Journal of Environmental Chemical Engineering* 11 (3):109719.

Tkaczyk, A. H., A. Bartl, A. Amato, V. Lapkovskis, and M. Petranikova. 2018. Sustainability evaluation of essential critical raw materials: Cobalt, niobium, tungsten and rare earth elements. *Journal of Physics D: Applied Physics* 51 (20):203001.

Usui, Takehiro. 2008. Estimating the effect of unit-based pricing in the presence of sample selection bias under Japanese Recycling Law. *Ecological Economics* 66 (2):282–288.

Valenta, Rick K., Éléonore Lèbre, Christian Antonio, et al. 2023. Decarbonisation to drive dramatic increase in mining waste–Options for reduction. *Resources, Conservation and Recycling* 190:106859.

Wath, Sushant B., Atul N. Vaidya, P. S. Dutt, and Tapan Chakrabarti. 2010. A roadmap for development of sustainable E-waste management system in India. *Science of the Total Environment* 409 (1):19–32.

White, Scott R., Benjamin J. Blaiszik, Sharlotte L. B. Kramer, Solar C. Olugebefola, Jeffrey S. Moore, and Nancy R. Sottos. 2011. Self-healing polymers and composites: Capsules, circulatory systems and chemistry allow materials to fix themselves. *American Scientist* 99 (5):392–399.

Xiaolong, Zhang, Zhang Shiyu, Liu Hui, and Zhao Yingliang. 2021. Disposal of mine tailings via geopolymerization. *Journal of Cleaner Production* 284:124756.

Yang, Jennie, Jens Bertram, Thomas Schettgen, et al. 2020. Arsenic burden in e-waste recycling workers – A cross-sectional study at the Agbogbloshie e-waste recycling site, Ghana. *Chemosphere* 261:127712.

Yin, Shi, Rabin Tuladhar, Feng Shi, Robert A. Shanks, Mark Combe, and Tony Collister. 2015. Mechanical reprocessing of polyolefin waste: A review. *Polymer Engineering & Science* 55 (12):2899–2909.

Yoshida, Aya. 2022. China's ban of imported recyclable waste and its impact on the waste plastic recycling industry in China and Taiwan. *Journal of Material Cycles and Waste Management* 24 (1):73–82.

Yvon, P., and F. Carré. 2009. Structural materials challenges for advanced reactor systems. *Journal of Nuclear Materials* 385 (2):217–222.

Zaidi, Syed Mohammed Javaid, Kenneth A. Mauritz, and Mohammad K. Hassan. 2019. Membrane surface nodification and functionalization. In *Functional Polymers*, edited by M. A. Jafar Mazumder, H. Sheardown, and A. Al-Ahmed. Springer International Publishing.

Zhuang, Wei-Qin, Jeffrey P. Fitts, Caroline M. Ajo-Franklin, Synthia Maes, Lisa Alvarez-Cohen, and Tom Hennebel. 2015. Recovery of critical metals using biometallurgy. *Current Opinion in Biotechnology* 33:327–335.

4 Understanding Metallurgical Processes and Wastes

Arunabh Meshram

4.1 INTRODUCTION

In the modern age of industrialization and rapid technological growth, the development of the metallurgical sector has also taken place in yesteryears. The demand for advanced alloys with specific applications has increased manifolds. Strategic and sophisticated applications dictate the demand on the specialized metallurgical industries. This enormous increase in the demand for specialized alloys has pressed metallurgical industries to produce and provide quality metals. The stringent quality measures of modern alloys have encouraged industries to achieve the required composition with minimum metal and material losses.

One of the most important metals in the service of mankind is iron. Steels (broadly, alloys of iron, carbon, and other elements like Cr, Mn, Ni etc.) are used in most of the important application areas, including structural and strategic areas. There are many classifications available for steels, based on compositions and applications. With the increase in the annual global production of steel, the generation of steel industrial wastes has also escalated. This includes the wastes generated at ironmaking and steelmaking sectors of the industries. These wastes (like flue dust, blast furnace slag, BOF slag etc.) are managed by the steel industry. In many industries, these wastes are discarded and landfilled, resulting in massive pileups. These pileups are responsible for the environmental pollution. Due to the presence of many hazardous elements, these wastes become the cause of groundwater pollution. The leaching of the heavy metals into the groundwater disturbs the pH and increases the hazardous metallic content. The flora and fauna around such massive pileups are affected drastically. The land and water pollution affects human health; these hazardous metals can cause life-threatening diseases, like cancer etc.

Similarly, aluminum is one of the most important metals used in many applications. With high strength-to-weight ratio, many lightweight components are developed using aluminum and its alloys. Aviation industry utilizes aluminum alloys for specialized applications. The classification of the aluminum alloys is based on the composition. The extraction of aluminum from its ore to produce metallic aluminum, followed by its alloying with other metals like magnesium, silicon etc. is a well-developed industrial process. Aluminum production industries have many processes that lead to the generation of metallurgical wastes (like Spent Pot Lining, Aluminum Dross, slags

DOI: 10.1201/9781003619680-4

etc.). These metallurgical wastes are subjected to metal recovery to some extent, however improvements in the processes can lead to reduction in the generation and proper recycling of the wastes.

An important issue that arises for metallurgical wastes is the necessity of modern technologies for efficient recycling. As mentioned previously, these wastes consume enormous land resources. The industries end up in selling and outsourcing these wastes to smaller industries that utilize the wastes for subsequent metal recovery or for development of products with lower metallic content. A large amount of waste is usually lost to the landfills, where it is discarded. The potential metallic content present in these wastes remains untapped as the recycling processes for these wastes are not available in every industry. Keeping the available resources as the constraint, industries have developed efficient recycling processes for the recovery of metallic values from these wastes, whereas researchers have explored modern laboratory technologies that could be translated to the industrial scale.

It is essential to understand that metallurgical waste recycling is key to materials cycle and its circularity. Material cycle is disrupted when metallurgical wastes are discarded and landfilled. Entrapment of metals in these wastes causes excessive environmental burden. It also is analogous to the economic loss incurred for losing metallic values. All recycled metals are reintroduced into the material cycle, where it is employed in the specific application till the end-of-life. Post application, the metallic values should be necessarily recycled to continue the material cycle.

The Chapters 4 and 5 cover the metallurgical industries and the wastes generated in them. These chapters will describe the overall management of metallurgical wastes, which are a segment of industrial wastes. In this chapter, the generation and recycling of metallurgical wastes will be elaborated extensively. Insights on the modern developments of recycling will be discussed. The following chapter is focused on extraction and beneficiation waste recycling and management. With these chapters, a holistic view of the metallurgical waste management is established.

In the next section, a broad classification of important metallurgical processes is provided. This classification also categorizes the extraction and processing of metals and alloys. With this classification, the categorization of the metallurgical wastes will also be provided, which are generated during these processes.

4.2 CLASSIFICATION OF METALLURGICAL PROCESSES AND INHERENT WASTE GENERATION

To understand the waste generation in metallurgical industries, it is essential to appreciate the different categories and classification of metallurgical processes. Broadly, the metallurgical production industries have been classified into three major categories, based on their processes: Pyrometallurgical, Hydrometallurgical and Electrometallurgical. This classification is done on the fundamental principles governing these processes.

As the name suggests, pyrometallurgical processes involve high-temperature operations like heating, melting, smelting. One of the most efficient routes of producing metals, pyrometallurgical process ensures complete utilization of the ores and generates the required products. A classic example of the same is iron ore smelting reduction by metallurgical coke in blast furnace. Ironmaking in the blast furnace is based on the smelting reduction process. Iron ores (hematite and magnetite) are charged in the blast furnace with appropriate fluxes and metallurgical coke. Hot blast gas is purged through the tuyeres to facilitate the smelting reduction. Metallurgical coke reacts with the iron ores to produce pig iron, which gets collected in the hearth. Countercurrent motion of the blast gas and the charged raw materials is unique in the blast furnace operation; it increases the contact of the blast gas with raw materials leading to better efficiency of metal recovery.

The metallurgical waste generated in the blast furnace smelting reduction process is the slag. There have been numerous research approaches to characterize and utilize the blast furnace and steelmaking slags. These slags have varying chemical composition, depending on the raw materials used and the process parameters followed during ironmaking and steelmaking. One of the major methods of utilizing blast furnace and Basic Oxygen Furnace (BOF) slags is in the field of making cement and other building materials. This has been put into practice by many industries as an efficient method of disposing the slag waste. However, better processes are being explored for implementation at the industrial scale.

The smelting process has been opted for many other metals like copper and lead. The generation of slags in these processes has led to the pileup of the waste in the industrial premises, as many industries lack the facility or process to efficiently dispose the wastes. Development of proper procedure for the utilization of the pyrometallurgical wastes like slags, salt cakes etc. essentially require the analysis and characterization of chemical composition of these wastes. After a comprehensive study of the composition, a process for a specific category of waste (like BOF slag) can be developed for a given composition range. With the segmenting of the pyrometallurgical wastes on the basis of the chemical composition, the waste characterization and utilization become more streamlined and efficient.

Hydrometallurgical processes involve the utilization of raw materials for extraction of metallic values into aqueous solutions, purification of these aqueous solutions and recovery of metallic values as metals, metal compounds, salts etc.

The key process governing the overall recovery of metallic value is leaching. The ores containing valuable ore-minerals of desired metals are subjected to comminution. After the conventional comminution, the ores are leached using a suitable chemical reagent. Leaching is the selective dissolution of raw material using aqueous solution of leaching reagent. These reagents can be acidic or alkaline in nature. Based on various leaching parameters like temperature, concentration of reagent, time, solid-liquid ratio, agitation, particle size etc. the recovery of metallic value into the leach liquor is achieved.

The leach liquor generated in the process is filtered to separate the residual solid and the pregnant leach liquor. The residual solids are subjected to further processing, based on their chemical analysis and characterization. These can be reintroduced for leaching or used as inert material for construction applications. The pregnant leach liquor is used for the metal recovery. It is possible that along with the metal of interest, other metals also get leached into the aqueous solution. The removal of these metallic values from the pregnant leach liquor is achieved in the purification stage of hydrometallurgical process.

The common methods of leach liquor purification include solvent extraction, ion exchange, precipitation, cementation and electrowinning. Based on the chemical composition of the leach liquor and the concentration of the impurity elements, these methods can be used. For example, to recover copper from the leach liquor, cementation, solvent extraction and electrowinning can be preferred. With multiple methods available, the process routes for hydrometallurgy are numerous [1].

After the recovery of most of the metallic values from the pregnant leach liquor, the leach liquor's chemical composition is determined. The acidic/alkaline concentration of the leach liquor is an essential parameter for this analysis. Raffinate solution (obtained after recovery of metallic values from pregnant leach liquor) can be subjected to further utilization as an input stream for the leaching process, as these solutions are fairly rich in ions.

The aqueous electrowinning electrolyte (which is produced after processing pregnant leach liquor), is used for recovery of metallic values by electrometallurgical process. Electrowinning cells dedicated for the purpose can be employed. Post recovery of the metal on to cathode, the spent electrolyte is reused in the process.

Most of the metals (non-ferrous) can be processed by hydrometallurgical route. The chief disadvantages of this route include the large volumes of leaching solutions and pregnant leach liquor, large investment for handling these solutions, chemical stream disposals and relatively lower metal recovery. As the temperatures of these processes are lower than pyrometallurgical processes, the rates of the reactions involved are lower. Also, as there are many alternative routes available, the metallurgical wastes produced are of varied chemical compositions and developing specific recycling route for the same is a challenging task.

Similar to the hydrometallurgical process, the electrometallurgical route of production is highly employed for metal recovery and refining. As noted above, in combination with the hydrometallurgical route, the electrowinning leads to the recovery of metal from aqueous electrolytes. For metals that are produced using fused salt electrolysis, the use of specific salt fluxes is essential. For example, Cryolite is used for aluminum extraction in Hall Heroult process. These molten salt fluxes are reused in the industry for maximum metal recovery.

Apart from the electrolysis processes for metal recovery, electrorefining is also carried out for achieving the desired degree of purity for specific applications. For instance, copper produced by a pyrometallurgical route can be subjected to electrorefining to remove the residual impurity elements that reduce its conductivity. The impure metal is taken as an anode; with a suitable electrolyte, the metal refining is achieved by supplying appropriate current. The refined metal is deposited at a cathode. In these electrometallurgical processes, the metallurgical wastes generated are essentially the spent electrolytes, spent electrodes or their parts and cell refractory linings.

For the three key metallurgical process routes described above, the number and the quantity of waste produced is enormous. Although most of the industries have developed methods of reusing waste streams, there are some metallurgical wastes that are released untreated. It is these wastes that cause environmental burden and are hazardous to flora and fauna. In the upcoming sections, the conventional routes of recycling metallurgical wastes in industries have been discussed. This will be followed by some modern advances in recycling and strategies to manage ever-growing waste. The significance of recycling metallurgical waste is also described to understand the effect on land and water resources, which in turn alters the environment. Finally, the chapter will be concluded with the emphasis on metallurgical waste management and its economic impact.

4.3 CONVENTIONAL MATERIAL RECYCLING IN METALLURGICAL INDUSTRIES

As mentioned in the previous section, the three major categories of metallurgical industries are pyrometallurgical, hydrometallurgical and electrometallurgical industries. In most of these industries, the waste is generated at every unit process. These wastes are recycled at the industries, if the chemical composition of the waste can be adjusted and the waste can be brought back into the material cycle by one of the unit processes present in the industry. In case the chemical composition of the metallurgical waste cannot be adjusted and requires significant addition of other raw materials, then these wastes are separated from the chief product and piled away. Physical/chemical properties of these wastes are such that reuse in unit processes is not possible. Red mud generated in aluminum industry in the Bayer's Process is one of the most important metallurgical wastes of this category. This waste is landfilled, resulting in huge pileups of untreated waste. The environmental impacts of improper waste disposal are discussed elsewhere.

One of the common materials used in all pyrometallurgical based industries is the refractories. Based on the extent of damage to the surface of the refractory material, newer bricks are added to the lining of the furnaces. These refractories are usually exposed to slag and hot molten metal. Due to the motion of these fluids at the surface of these refractories, erosion and chemical damage are observed. These refractory bricks are recycled and the furnaces are relined, after discarding/removing the damaged surface.

A common example of this reuse is the recycling of spent pot lining (SPL). SPL is the waste lining generated at the smelter unit of aluminum industry. Hall Heroult cell uses a cathode pot that is lined with refractories. SPL produces two cuts: refractory cut and carbon cut. The refractory cut is directly reused in the next pot lining, as it is not exposed directly and therefore, does not experience much damage. The carbon cut is rich in fluoride, aluminum and other ionic species and it is subjected to recycling. One of the methods being explored in laboratories is to use the carbon cut as a raw material for hydrometallurgy after comminution.

Another most common path of metal recycling in metallurgical industries is remelting of casting, foundry and manufacturing defects. Due to various manufacturing defects observed at various stages of production, large quantities of in-house discards are generated. Some important defects are observed at casting (continuous, ingot and intricate shape casting at foundry shops) and mechanical processes like

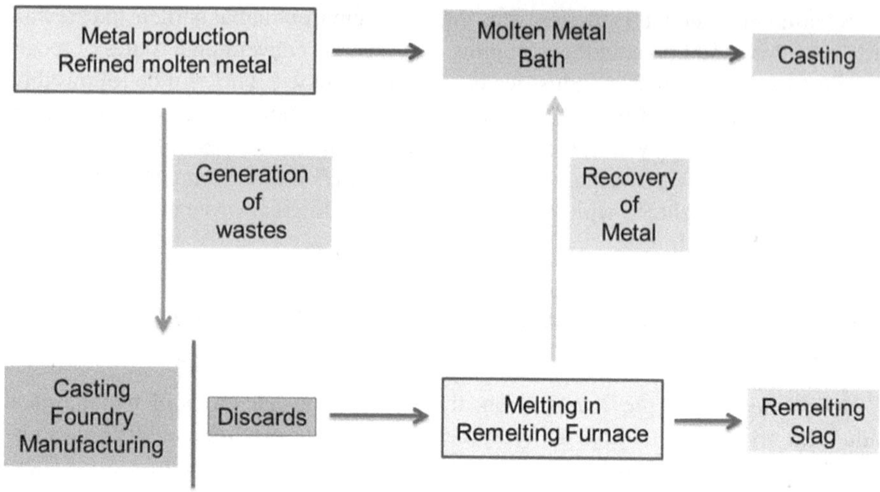

FIGURE 4.1 A schematic diagram to understand the recovery of metal from various casting and manufacturing rejects. Generation of slag at the remelting stage also requires additional processing to ensure complete metal recovery.

forging, rolling, extrusion, drawing etc. Defects like cold shuts are noted due to improper manufacturing operations.

For the efficient metal recovery, there are remelting units that can have melting furnaces. These furnaces are used for recovering the metal into the material cycle. Figure 4.1 illustrates the path of these wastes.

The molten metal in the remelting furnace can be reintroduced into the molten metal bath upon composition adjustment, if required. The presence of other metals and trace elements and gangue can require additional refining, before subjecting to application. This ensures the efficient utilization of the metal from such wastes. During the melting at the remelting unit, the slag generated also will have some metal loss entrapped in the slag matrix. For complete metal recovery, the recycling of these remelting slags is also essential.

Slags are metallurgical wastes generated in pyrometallurgical processes. Smelting reduction, melting and remelting of ingots lead to generation of slags. Slags are essential for metallurgical processes as they serve as thermal barrier for molten metal (reducing the thermal energy loss to the environment), physical barrier (protecting the molten metal from oxidation and other reactions) and sink for impurities (emerging from the molten metal). Slags are produced by adding fluxes to the molten metal. Slags have relatively low density compared to the molten metal; therefore they float above the molten metal.

One of the key objectives of modern metallurgical processes is to reduce the slag production. It has been reported that good stirring of the molten bath ensures low phosphorus concentration in steel. Similarly, optimizing blowing operation can lead to steady increase in low P grade and low N grade steels [2].

As noted by Cui et al., slags can be used as a raw material for the development of stone mosaic asphalt. Some important properties of steel slag that govern the use of asphalt production are abrasion resistance, polishing resistance and the inherent morphology of the slag aggregates [3]. Also, for basic oxygen furnace (BOF) steelmaking slag, the amount of heat energy contribution to the heat balance is nearly 269 MJ, which is around 11.7% of total energy calculated for operation of BOF [4]. Therefore, it is crucial to understand and appreciate energy balance along with the material flow involved in metallurgical processes. A holistic recycling process should cater to the recovery of waste energy as well.

Application of steelmaking slag for direct use has been examined by many researchers. According to Mutsuda et al., the ocean environments can be restored by using steelmaking slags. A numerical method using the Eulerian-Lagrangian model developed by Mutsuda et al. has helped in concluding that a metallurgical waste (steelmaking slag) can be successfully used for reducing dissolved sulfide concentrations [5].

Using in-site XRD analysis, Chai et al. have proposed the utilization of steelmaking slags for super and subcritical carbonation reactions for generation of calcite and aragonite [6]. The carbonation of these slags under pressurized systems has led to the phase transformation of Ca-bearing phases to valuable mineral phases. A similar method has been explored by Ma et al.; they have used powdered slags samples to produce minerals by dissolving in water and introducing CO_2 into the solution [7]. The mineralization has been confirmed using TGA-DTA analysis. It has been reported that the presence of $Ca(OH)_2$ in steelmaking slag reacted with CO_2.

To develop a holistic approach to recycle slags, a combination of physical treatments like crushing, grinding and mineral beneficiation is done to enhance the metal content. Liu et al. have used the physical treatment operations to enrich the metallic content from the slags [8]. Coarser particles fraction (richer in iron content) was returned to smelting operation to recycle Fe into material cycle, whereas finer particle fraction was used for cement production.

Table 4.1 describes some important wastes generated in metallurgical industries. As noted in the table, a broad classification of ferrous and non-ferrous wastes has been presented. The conventional methods of reutilization of these wastes have also been described. Some important developments regarding waste energy utilization are also presented.

As observed in Table 4.1, there are many categories of ferrous and non-ferrous metallurgical wastes. These wastes are mostly utilized for developing valuable materials, if extraction of metal directly is not economically feasible. Subjecting these wastes to melting operation for extraction of metal is a common practice. White aluminum dross is rich in metallic aluminum; therefore its melting to liberate molten aluminum is economically feasible. The conventional method of recovering metallic aluminum is to remelt the dross in a rotary salt furnace in the presence of salt fluxes. This leads to the separation of molten aluminum and salt slag (called salt cakes) [56, 57]. Apart from the conventional routes of remelting the dross in RSF, there are newer salt free technologies that employ plasma arc technologies, electric energy sources for melting the dross with negligible salt flux addition [56, 58].

TABLE 4.1
Metallurgical wastes and some important conventional methods of recycling

Broad class of waste	Type of metallurgical waste	Methodology/ Key processes involved	Recycling products	Reference
Ferrous	Blast furnace slag	Used as a raw material for concrete mixture	Concrete	[9–11]
		GGBFS (47 wt.%) used for concrete mixture	Concrete	[12]
		GGBFS (50 wt.%) used for concrete mixture	Recycled concrete aggregate	[13]
		GGBFS (25 wt.%) used for concrete mixture for studying global warming potential	Recycled concrete aggregate	[14]
		GGBFS (20–40 wt.%) used for concrete mixture for high temperature applications	Recycled concrete aggregate	[15]
		GGBFS (5–25 wt.%) used for cement composite production	Lightweight hemihydrate phosphogypsum-cement composite	[16]
	Basic oxygen furnace (BOF) slag	Steelmaking slag leached with alkaline and acidic solutions	Phosphate precipitates	[17, 18]
		Steelmaking slag used as fertilizer for wheat growth	Fertilizers for agriculture	[19, 20]
		Waste heat recovery from slag for pyrolysis of coal and municipal wastes	Heat recovery	[21]
		Use of BOF slag as source of Fe(II) ions in water	Fe(II) ions in water	[22]
		BOF slag and phosphate tailings used for recovery of P by leaching	Phosphorous recovery as phosphate fertilizer	[17, 23, 24]
	Electric arc furnace (EAF) slag	EAF slag used as raw material for concrete production	Alkali-activated concrete	[25–30]
		Use of EAF slag as source of heavy metals for lipid synthesis by leaching and culture medium preparation	Lipid synthesis of C. pyrenoidosa	[31]
		EAF slag as a raw material for asphalt mixture	Semi-dense asphalt, asphalt mixtures	[32, 33]
		EAF slag as raw material for concrete production. Radioactive shielding performance of concrete determined	Concrete	[34]
Non-Ferrous	Copper smelter slag	Copper smelter slag subjected to direct-reduction with Na_2CO_3 and CaO addition and magnetic separation	Iron and copper	[35]

	Copper smelter slag subjected to smelting and separation	Iron	[36]
	Reduction of copper smelter slag	Iron-rich alloy containing Cu and Mo and secondary slag	[37]
	Copper smelter slag subjected to oxidation, hydrothermal treatment, filtration, gel formation	Silica gel	[38]
Zinc smelting slag	Pb-Zn smelting slag combined with alkali activators and fly ash and cured	Alkali-activated cementitious material	[39]
	Brass smelter slag (Cu-Zn rich) leached with $ZnCl_4$-NH_4Cl and purified	Copper and Zinc	[40]
	Zinc smelter slag leached, followed by precipitation	Cobalt and Zinc	[41]
	Pb-Zn smelting slag used for wastewater treatment	Ferric arsenate precipitate	[42, 43]
	Zn slag used for treating copper smelter wastewater	Arsenic precipitation	[42, 44]
	Zn slag and clay used for brick making by oxidation sintering	Lightweight bricks	[45]
Lead smelter slag	Lead smelting slag combined with other solid wastes by grinding, mixing, and curing	Mine cemented backfill material	[46]
	Variation of lead oxide in slag mortar composition to achieve required strength	Alkaline activated slag mortar	[47, 48]
	Lead-zinc smelting slag, electrolytic manganese residue and blast furnace slag mixed, rolled and cured	Geopolymers	[49]
	Lead smelter slag, waste glass sand, fly ash, blast furnace slag mixed with binders and molded	Geopolymers	[50]
	Lead slag subjected to solidification/ stabilization using microwave technology	Geopolymers	[51]
	Lead-zinc tailings melted, quenched and ball milled and sintered	Glass ceramics	[52]
	Municipality solid waste incineration fly ash blended, poured, and molded with lead slag	Magnesium potassium phosphate cement	[53]
	Lead slag subjected to carbothermic reduction and thermal volatilization	Zinc, lead and tin	[54]
	Lead slag ground and molded by grouting molding method	Lead smelter slag based cementitious material	[55]

Dross is a metallurgical waste generated at melting/remelting unit or in any process where molten metal comes in contact with atmospheric oxygen. The key feature of dross is the presence of the oxide phase that entraps molten metal in its matrix. For instance, aluminum dross is formed when molten aluminum comes in contact with oxygen. The top surface of molten metal gets oxidized and simultaneous entrapment of aluminum takes place. This semisolid skim is removed before casting the molten aluminum. On cooling, the semisolid skim gets further oxidized and solidified into a hard mass called dross. The dross can have a different composition, depending on the composition of the source (metal or alloy) and the process leading to its generation [59, 60]. It is necessary to note that apart from the oxide and the metal entrapped in the oxide matrix, there are other metals present in relatively lower concentration. The recycling of dross also ensures the recovery of these metals into the material cycle.

Aluminum dross has been used as a raw material to develop valuable materials using hydrometallurgical route of recycling. Important materials like potash alum, tamarugite, alumina are produced by leaching aluminum dross with acidic and alkaline solutions [61, 62]. When aluminum dross comes in contact with water, most of the dissolvable impurities are leached into the water. However, if an alkaline solution is used for leaching, the development of important gases like H_2 CH_4 and NH_3 has been reported in literature. On examining the samples of dross it has been put forth that the composition of dross is the governing factor for the evolution of these gases. Aluminum carbide and nitride are the phases responsible for the generation of ammonia and methane, whereas hydrogen evolution is governed by the reaction of aluminum and water in alkaline conditions [60, 62, 63]. Important metals like aluminum, zinc, and lead, exhibit dross generation.

Wastes generated in copper, lead, and zinc industries have also been discussed in Table 4.1. After proper neutralization, these slags can be used for construction applications. The presence of these metals in the slags impacts the environment. The leaching tendency of these metals into groundwater is considered a hazardous effect of these slags. A stabilized application of these wastes for construction applications (cement, concrete, and other building materials) is a relatively safer method of waste utilization, compared to landfilling and generation of pileups.

For the metals that are produced by the electrometallurgical route, the generation of electrolyte waste is evident during the electrowinning or fused salt electrolysis process. In the electrowinning process, the metal is deposited from the aqueous solution (generated after purification of the leach liquor by solvent extraction or other method). After the deposition of metal from the electrolyte, the reutilization of the electrolyte is done. It is because the electrolyte is highly acidic in nature. Reutilization of the electrolyte ensures efficient ionic recovery.

Other sources of spent electrolytes are the plating and electrorefining processes. The electrometallurgical process where the impure metal from anode is dissolved into the electrolyte and simultaneously deposited on a pure cathode is electrorefining.

Pregnant leach liquor is produced after leaching lean grade ores with suitable reagents. Process parameters like leaching time, temperature, agitation, solid-liquid ratio and concentration of reagents are optimized to achieve high leaching efficiency. The leach liquor rich in metallic content is subjected to purification by solvent extraction or ion exchange processes. These processes ensure metal recovery in other solutions. The leach liquor is used to extract the metallic values into these stripping solutions through organic solvents or resin beads, as applicable. Raffinate solutions are formed after metallic content is extracted from the leach liquor.

Similar to electrolytes used in the electrowinning processes, the raffinate solutions are rich in ions (cations and anions) as these solutions are produced from leach liquors, which in turn, are produced using acids and alkalis. The utilization of the raffinate solutions has been followed in industries by adjusting the concentration of the ions of the solution. Recirculation and reuse in the production cycle as an additive to the fresh leaching reagent is one of the important methods of employing the raffinate solutions.

The next section describes the advances in metallurgical waste management. The newer processes involving the utilization of metallic values from these wastes for generation of valuable products have been discussed further. It is important to note that many processes with promising results are still being investigated for scaling up to meet the industrial needs.

4.4 DEVELOPMENT OF REUSE AND RECOVERY PROCESSES FOR METALLURGICAL WASTES

One of the key factors in governing the development of recovery processes for metallurgical wastes is to understand the importance of waste generation at every step during the process. For instance, as shown in Figure 4.2, the beginning of metallurgical waste generation is marked at the initial comminution step, followed by secondary treatment of ores, further channeling into pyrometallurgical, hydrometallurgical or electrometallurgical processing. The refining of the products is also taken into account; the casting of the metal into desired shapes is the final step involved. A wide variety of wastes are generated throughout the process, like primary and secondary slags, spent raffinate solutions, electrolytes, and electrolytic cell wastes. As described in the previous section, these wastes are either rich in metallic value or are hazardous to the environment. Development of proper methods for utilization of these wastes is necessary. Important processes involved in mineral beneficiation and the inherent waste generated will be discussed in Chapter 5. Some key wastes that require further processing include the beneficiation wastes generated at these sites.

Although many conventional processes are present for the reutilization of these wastes within the industries, there are many research ventures for the development of newer technologies for effective metal reutilization. These newer methods are being investigated at laboratory stages. This section will discuss the modern technological advances in the field of metallurgical waste recycling. For instance, Table 4.2 describes the composition of slags discussed in the previous section and the advances in the recycling processes.

There are numerous other metallurgical wastes produced in iron and steelmaking industries. Dust generated in the blast furnace and basic oxygen steelmaking process is usually rich in zinc. The production of pig iron nuggets has been examined

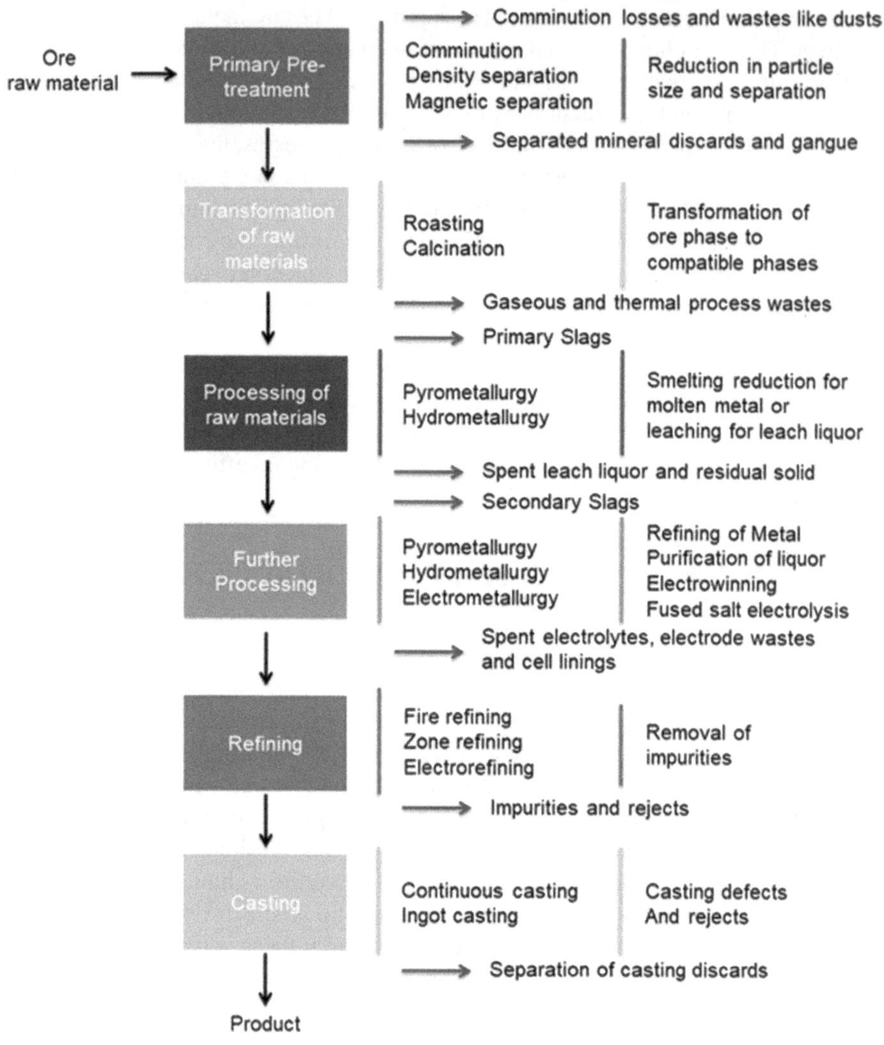

FIGURE 4.2 A schematic diagram to understand the holistic perspective of metallurgical waste generation.

and its thermodynamic feasibility has been validated at ~1450 °C [72]. Zinc can be recovered as oxide and recycled into the material cycle.

Sukmak et al. have discussed the possibility of using EAF slag as raw material for construction applications. Their investigation focused on the reaction of EAF slag with water and an acetic acid environment. It has been reported that volume expansion is observed for water exposure, whereas dissolution of Ca-phases led to a reduction in volume for acidic environment [73]. Kim and Azimi have applied carbothermic reduction of EAF slag for recovery of metals like Fe, Mn, Cr and Nb in metal phase

TABLE 4.2
Modern advances in slag recycling

Type of slag	Technological advances in laboratories	Resultant Product/ key outcomes	Reference
Basic oxygen furnace slag	Recycling of BOF slag fines using physical separation using magnets	Separation of BOF slag fines for application in different sectors	[64]
	Use of extractive grinding process to enhance Ca-recovery from BOF slag	$CaCO_3$/PCC by selective dissolution of slag fines	[65, 66]
	Silica modification of BOF slag by oxidation for recovery of magnetic iron oxide	$(Mg,Fe^{2+})O.Fe_2O_3$ (to be used for magnetic applications)	[67]
	Use of BOF slag as dephosphorization agent	~20% replacement of dephosphorization agent	[68]
Electric arc furnace slag	Used for making ceramic tile by mixing, drying and hydraulic pressing	Ceramic tile	[69]
	EAF slag as raw material for concrete beams (cylindrical)	Concrete beams	[70]
	EAF slag fired at high temperature and shaped	High-density ceramics	[71]

and Ti, Mg and Al in slag, followed by acid baking-water leaching to recover metals from slag phase [74]. This process described the metal recovery by employing multi-stage operations for complete metal recovery.

Similarly, the recycling of copper smelter slag has been investigated by many researchers. Copper smelter slag is generated during the copper matte smelting. As it is rich in Fe, Cu, iron oxides and other phases, the recycling of the CSS can lead to the generation of valuable materials [36]. Flash smelting furnace is another source of copper slag, which requires additional processing route for its recycling. As described by Huihui Zhou et al., the presence of arsenic in the copper slag is due to the copper-arsenic alloys [75]. A partition of arsenic is observed for concentrate and tailing fractions. In tailings, arsenic is present as silicate, whereas in concentrates, it is associated with copper sulfides, which makes its separation difficult.

It has been observed that smaller particle size of smelter slag implies higher surface area for contact. This also leads to a higher probability of oxidation and wearing in such smaller particles. The presence of Cd, Cu, Mn, Pb, and Zn in zinc slag leads to higher environmental burden, as these elements are potentially hazardous [76, 77]. Zinc leach residues (ZLR) have been used as raw material for recovery of Zn and Pb by rotary kiln operation. Addition of coke powder and calcium oxide is done [78].

With an insight into modern metallurgical waste processes, it is necessary to appreciate the importance of metallurgical waste recycling on a holistic scale. The next section elaborates the significance of recycling these wastes with an emphasis on the impact on natural resources due to improper waste disposal.

4.5 SIGNIFICANCE OF METALLURGICAL WASTE RECYCLING

Every metallurgical process has a defined set of inputs and outputs. The raw materials of a process can help in ascertaining the properties of the finished product. This analogy also applies to the waste generated in these processes. The metallurgical waste produced can also be subjected to evaluation to determine the losses of metallic values in them. When the estimation of the metallic values is done for a given waste category, the next step is to adopt an efficient recycling route.

Analysis of the metallurgical waste from pyrometallurgical, hydrometallurgical or electrometallurgical route provides crucial statistical data about the waste. Development of an appropriate process becomes crucial for achieving the desired metal recovery. The metallurgical waste generated in the industries ends up in pileups and it causes many consequences, including land and water resource pollution and environmental burdening.

To understand the extent of land resource exploitation, an example of red mud is considered. Red mud is produced in aluminum production industries. A metallurgical waste produced during the leaching of bauxite ore, red mud is piled and stacked in large landfill ponds. For every ton of aluminum metal produced, nearly one ton of red mud is generated. This implies that a sharp need for developing a recycling method is necessary, as red mud is hazardous due to its high alkalinity and presence of heavy metals. Presently more than 3 billion tons of red mud is stored in ponds and dried pits, making the land resources around such pits unfit for other applications like cultivation etc.

The presence of hazardous materials in these metallurgical wastes can lead to disruption in the flora and fauna of the environment. For example, CaO is present in most of the steelmaking slags. Ironmaking and steelmaking slags are discarded in the slag pits in many steelmaking industries. Although recent advances indicate the utilization of slags for cement and brick making applications, the slags are landfilled to a large extent. Oxides of iron, calcium, silicon, and many other phases lead to change in pH of the surrounding soil.

Technogenic magnetic particles, polycyclic aromatic hydrocarbons (PAH) and heavy metals have been detected in soil around areas where coke plants and metallurgical industries were located [79]. Coking plants contribute to the emissions that lead to an increase in PAH content in the soil. Heavy metals like Cd and Pb are detected due to the emissions of metallurgical industries. A similar investigation has been done for Panzhiuha, a city in China by Wang et al. [80]. It has been observed that the ecological risk index is highest for industrial areas, followed by heavy traffic areas and lowest in residential areas. Pb, Zn, Cu and Cr have higher concentration in the industrial dust, compared to the local soil. These metals contribute to the degradation of the land resource.

The effects of ore mining and processing industry in southern Poland on the local environment have also been investigated [81]. With mining and processing of Zn-Pb ores in vicinity, the concentrations of Zn and Pb in agricultural soil are higher than in other local areas. It has been suggested that agricultural activities should be excluded and technological reforms should be done to make the soil reusable. A similar examination of coastal areas of Bangladesh has been done [82]. Nearly 12% contribution of heavy metals/metalloids is due to metallurgical industries and 23% wastes are directed from industrial effluents. Awareness should be developed extensively to ensure higher metal recycling and lower land and water pollution.

Analysis of copper industrial waste generation indicates that phases like pentlandite, pyrotite, chalcopyrite, chalcite, tenorite etc. are present in the metallurgical dust along with oxides of As and Pb. The recycling of the dust is a challenge for copper industries [83].

Water resources are also deeply affected due to the metallurgical wastes. Most of these wastes, as mentioned above, are discarded in the landfill pits and ponds. This landfilling also leads to pollution of groundwater. The pH of the groundwater changes from its natural value to alkaline (in the presence of highly alkaline wastes) or acidic (due to acidic wastes). This hampers the growth of plants, subsequently affecting the food chain and environmental life cycle.

A different perspective is seen when wastewater produced in the steelmaking industry is studied. Nearly 28.6 m^3 water is required for production of 1 ton of steel. It contains primarily of many metals and hydrocarbons (due to various processes in the steelmaking industry, like mining, smelting, steelmaking at LD converter and rolling mills). Many methods of wastewater recycling have been discussed in the modern literature, including use of different precipitating methods like organic solvent settling coupled with vacuum filtration, bacterial recycling (to recover Fe from the wastewater) and photocatalysis to degrade the CN^- ions in the wastewater.

Ji et al. have studied the utilization of metallurgical slag for wastewater treatment [84]. Some of the important methods for utilization of slag included the precipitation of arsenic from wastewater (as slags contain abundant Ca and Al) and removal of antibiotics like Norfloxacin using particle electrodes comprising of steel slag, $NaHCO_3$ and Kaolin. These studies indicate that metallurgical wastes are being explored as raw materials for water recycling.

Mele and Magazzino have employed machine learning to understand the connection between the industrial waste generation, air pollution and economics in China [85]. It has been reported that over the time of past decades, efforts towards sustainable economic growth have led to a reduction in environment pollution. This can also be attributed to the increase in education of the public in environmental sciences.

Similarly, the waste generation, including CO_2 emissions and energy requirements for nickel production industry have been studied by Fukuzawa [86]. With the advent of modern regulations on the supply end of industry, the industries are focused on developing cleaner production routes. Social awareness and comprehensive economic model encompassing the demand and supply end of metals could make the production more sustainable.

With the growing insight on metallurgical waste generation and its environmental impact, the LCA of metallurgical wastes have been done by Di Maria et al. [87]. These studies have shown that LCA can be taken as a method of determining chief opportunities and benefits of a circular economy to recycle these wastes. These LCA reports can also provide alternative methods to design efficient technology aiming for economic and environmental benefit, while exploring the key drawbacks of the existing recycling processes.

Not all recycling processes lead to environmental benefit. The LCA should be done such that the environmental impact of recycling of such wastes is also determined. Acharya et al. have used ferrochrome ash waste to generate blended concrete and found that this blended concrete had a relatively lower environmental impact compared to normal concrete [88].

Appreciation of natural resource utilization during metal production is essential. Similarly, it is necessary to understand the impact of metal production and its side-effects like waste generation on the environment. These aspects have been elaborated above. In the next section, the management of wastes generated in metallurgical industries will be discussed.

4.6 METALLURGICAL WASTE MANAGEMENT

With the growing industries of the modern world, newer applications and uses are being developed. This growing trend is also observed in the field of metallurgical industries. For instance, advanced strategic applications for defense equipment have been developed using non-ferrous metals and rare-earth elements. Proper extraction, purification and application of these elements have become crucial for these metallurgical industries.

To appreciate the importance of metallurgical waste management, it is essential to study the material flow in the wastes. For instance, aluminum dross is a metallurgical waste from an aluminum casting unit. Depending on the type of dross, the concentration of aluminum in the dross could be as high as 75–80 wt.%, making dross an important material containing metallic aluminum [89]. Therefore, a definite route of metallurgical waste recycling is required for dross to recover aluminum.

The possibility of establishing a method for metal extraction from wastes has been studied and implemented in many industries. In this section, metal recycling within the industries will be discussed. This will be followed by understanding the importance of metallurgical waste management and its impact on human society.

As discussed previously, copper raffinate is generated after the extraction of copper from the pregnant leach liquor in the solvent extraction process. As this solution is rich in ions (primarily due to the high acid concentration), the reutilization of the raffinate solution has been implemented on a large scale. Similarly, spent copper electrolyte solution, produced in the electrowinning process of copper production, is also highly concentrated with acids [90]. These solutions are recycled back into copper leaching/refining stages, after chemical composition adjustment. Taking the example of zinc ash, a metallurgical waste produced in a hot dip galvanization unit.

Nearly 60–90 wt.% Zn is present in zinc ash. As the ash is rich in zinc, the recovery process aims to extract maximum metal from it [91, 92].

One of the most important objectives of metallurgical waste management is to achieve metal reclamation within the economic constraints, while investing minimum resources. The quality of metal recovered can be improved, if necessary, by refining. Most of the industries intend to reclaim the metallic waste by subjecting it to remelting and casting process. This leads to recovery of most of the metallic fraction; however secondary casting wastes are produced in such cases. Also, complete metal recovery is not ensured, as metal tends to get entrapped in the slag phases. This indicates the necessity of developing another waste management process for these wastes.

The economic constraints present in metal recycling processes that govern the feasibility of the process are chiefly categorized as capital expenditure and operation expenditure. Usually, the development of a separate waste management process completely dedicated to recycling a metallurgical waste may not be present in small scale metallurgical industry. The capital investment for such processes could outweigh the output of recycling the wastes. A more economical alternative is to sell/transfer the waste to another industry that has such a facility. Many smaller steelmaking industries follow this to dispose steelmaking wastes like slags etc.

Development of a metal recycling unit in an established industry would require additional investment in the direction. Such investment, as described above, involves the capital expenditure and operational expenditure. On the other hand, provisions should be made during the process of the establishment of the industry. Reserving land resources, energy resources, water and essential infrastructure should be included in the planning stage of establishing the industry. This can help in understanding the material flow within the industry and improve the logistics involved. Also, the wastes generated in the industries can be efficiently classified and sorted, provided the facility for temporary storage is available. These wastes can then by subjected to further processing, as described in the previous sections.

One of the important concepts applied to industrial waste management is the evaluation of the crucial factors involved in material flow and economics of the process. To begin the waste management process, the chief factors are society, use of waste materials, methods of recycling, inherent economics, recovery technology, manufacturing, design of finished product and environmental impact [93]. To recycle wastes generated in industries, comminution has always been a key process for dismantling the material and separating various components into metallic and non-metallic fractions.

The generation of waste from various sources can be reduced by developing newer products with lesser environmental impacts. Minimization of waste depends on how existing products are manufactured. The waste storage and disposal are connected to the availability of land resources for safe accumulation. As land resources available for waste disposal are scarce and soil pollution is evident in landfills, storage, and landfills are discouraged for longer periods. As described by Farzadkia et al., the minimization of waste should be implemented on a larger scale [94]. For the disposal of industrial waste, there are various regulations as observed by the government of different countries.

Srivastava et al. have explored the case studies of industrial waste management of red mud. Based on the physical characteristics and hazardous impact of the red mud on the environment, a wide range of developed processes have been studied. It has been noted that red mud can be used as a raw material for blending in construction applications. Metal recovery (Al, Sc, Ti and other REE) from red mud by hydrometallurgical operations is also reported [95].

To conclude, it is necessary to understand that metallurgical wastes are good sources of metallic values. Designing a feasible method for metal recovery or production of valuable material from these wastes should be one of the key aspects of these metallurgical industries. This will ensure higher metal circularity and higher metal recovery efficiency. This chapter has linked the metallurgical processes to waste generation and industrial and advanced methods of recycling. Environmental and economic aspects of designing a metallurgical waste processing method have also been described.

As has been described previously, based on the type of industry, pyrometallurgical, hydrometallurgical or electrometallurgical, the wastes are generated. Important wastes like slags, dross, raffinate etc. have been described to enhance the understanding of the variety of wastes produced in these industries. Development of efficient methods of metal recovery should be designed keeping the economics of establishing separate recycling facility in the industries. The waste management and its importance have also been illustrated.

The improvement in metal recovery implies higher production. An overall outlook on the metal recovery is essential for developing such recycling processes. Many efficient recycling routes are being explored recently. Scaling these processes to meet the industrial demands will also increase the recycling of these wastes.

REFERENCES

[1] Ray HS, Sridhar R, Abraham KP. *Extraction of nonferrous metals* (2011th ed). New Delhi: Affiliated East-West Press Pvt. Ltd., 1985.

[2] Zhao C, Sun Y, Tian Z, et al. Practice of hot metal full pretreatment and less-slag steel making in Shou Gang Jing Tang. *J Iron Steel Res Int* 2014; 21: 104–107.

[3] Cui P, Ma T, Wu S, et al. Texture characteristic and its enhancement mechanism in stone mastic asphalt incorporating steel slag. *Constr Build Mater* 2023; 369: 130440.

[4] Kumar B, Roy GG, Sen PK. Comparative exergy analysis between rotary hearth furnace-electric arc furnace and blast furnace-basic oxygen furnace steelmaking routes. *Energy Clim Chang* 2020; 1: 100016.

[5] Mutsuda H, Miyata Y, Doi Y, et al. Numerical investigation into the restoration of ocean environments using steelmaking slag. *Mar Pollut Bull* 2018; 131: 428–440.

[6] Chai YE, Miller QRS, Schaef HT, et al. Pressurized in situ X-ray diffraction insights into super/subcritical carbonation reaction pathways of steelmaking slags and constituent silicate minerals. *J Supercrit Fluids* 2021; 171: 105191.

[7] Ma Z, Liao H, Cheng F. Synergistic mechanisms of steelmaking slag coupled with carbide slag for CO 2 mineralization. *Int J Greenh Gas Control* 2021; 105: 103229.

[8] Liu X, Gao P, Han Y. Resource utilization of slag from desulphurization and slag skimming : A comprehensive recycling process of all components. *Int J Min Sci Technol* 2022; 32: 585–593.
[9] Majhi RK, Nayak AN. Production of sustainable concrete utilising high-volume blast furnace slag and recycled aggregate with lime activator. *J Clean Prod* 2020; 255: 120188.
[10] Majhi RK, Nayak AN, Mukharjee BB. Characterization of lime activated recycled aggregate concrete with high-volume ground granulated blast furnace slag. *Constr Build Mater* 2020; 259: 119882.
[11] Abhishek P, Ramachandra P, Niranjan PS. Use of recycled concrete aggregate and granulated blast furnace slag in self-compacting concrete. *Mater Today Proc* 2021; 42: 479–486.
[12] Gholampour A, Zheng J, Ozbakkaloglu T. Development of waste-based concretes containing foundry sand, recycled fine aggregate, ground granulated blast furnace slag and fly ash. *Constr Build Mater* 2021; 267: 121004.
[13] Biswal US, Dinakar P. A mix design procedure for fly ash and ground granulated blast furnace slag based treated recycled aggregate concrete. *Clean Eng Technol* 2021; 5: 100314.
[14] Shamass R, Rispoli O, Limbachiya V, et al. Case studies in construction materials mechanical and GWP assessment of concrete using blast furnace slag, silica fume and recycled aggregate. *Case Stud Constr Mater* 2023; 18: e02164.
[15] Tung TM, Babalola OE, Le D. Experimental investigation of the performance of ground granulated blast furnace slag blended recycled aggregate concrete exposed to elevated temperatures. *Clean Waste Syst* 2023; 4: 100069.
[16] Zhang L, Hung K, Poh S, et al. Mechanical strength, water resistance and drying shrinkage of lightweight hemihydrate phosphogypsum-cement composite with ground granulated blast furnace slag and recycled waste glass. *Constr Build Mater* 2022; 345: 128232.
[17] Yu Y, Du C. A review on the P enrichment and recovery from steelmaking slag : Towards a sustainable P supply and comprehensive utilization of industrial solid wastes. *Sci Total Environ* 2023; 891: 164578.
[18] Yu Y, Du C, Fan S, et al. Acid-leaching separation of phosphorus from the BOF slag modified with Al 2 O 3. *J Environ Chem Eng* 2022; 10: 108394.
[19] Fisher LV, Barron AR. Effect of functionalized and unfunctionalized basic oxygen steelmaking slag on the growth of cereal wheat (Triticum aestivum). *Resour Conserv Recycl Adv* 2022; 15: 200092.
[20] Fisher LV, Barron AR. The recycling and reuse of steelmaking slags — A review. *Resour Conserv Recycl* 2019; 146: 244–255.
[21] Zhang H, Dong J, Wei C, et al. Future trend of terminal energy conservation in steelmaking plant : Integration of molten slag heat recovery-combustible gas preparation from waste plastics and CO2 emission reduction. *Energy* 2022; 239: 122543.
[22] Yu D, Liu F, Zou C, et al. Municipal green waste promotes iron release from steelmaking slag in water. *Resour Conserv Recycl* 2023; 188: 106722.
[23] Yu Y, Du C, Yang X. Recovery of phosphorus from steelmaking slag and phosphate tailings by a collaborative processing method. *Sep Purif Technol* 2023; 313: 123499.
[24] Yu H, Lu X, Miki T, et al. Sustainable phosphorus supply by phosphorus recovery from steelmaking slag : A critical review. *Resour Conserv Recycl* 2022; 180: 106203.
[25] Hafez H, Kassim D, Kurda R, et al. Assessing the sustainability potential of alkali-activated concrete from electric arc furnace slag using the ECO 2 framework. *Constr Build Mater* 2021; 281: 122559.

[26] Alsheltat MF, Abdulmalik Elfigih M. Effects of electric arc furnace slag powder and fly ash within ternary waste blend on performance of concrete. *Open Ceram* 2023; 14: 100359.
[27] Amani A, Ramezanianpour AM, Palassi M. Investigation on the sustainable use of electric arc furnace slag aggregates in eco-friendly alkali-activated low fineness slag concrete as a green construction composite. *J Clean Prod* 2021; 307: 127257.
[28] Li Y, Qiao C, Ni W. Green concrete with ground granulated blast-furnace slag activated by desulfurization gypsum and electric arc furnace reducing slag. *J Clean Prod* 2020; 269: 122212.
[29] Sosa I, Thomas C, Polanco JA, et al. Durability of high-performance self-compacted concrete using electric arc furnace slag aggregate and cupola slag powder. Cement and Concrete Composites 2022; 127, 104399. DOI: 10.1016/j.cemconcomp.2021.104399
[30] Ozalp F. Effects of electric arc furnace (EAF) slags on mechanical and permeability properties of paving stone, kerb and concrete pipes. Construction and Building Materials 2022; 329, 127159. DOI: 10.1016/j.conbuildmat.2022.127159
[31] Yu Q, Liu T, Zeng Y, et al. Efficient lipid synthesis of Chlorella pyrenoidosa promoted under heavy metals from electric arc furnace slag. *J Clean Prod* 2023; 414: 137648.
[32] Mikhailenko P, Piao Z, Poulikakos LD. Case studies in construction materials electric arc furnace slag as aggregates in semi-dense asphalt. *Case Stud Constr Mater* 2023; 18: e02049.
[33] Ali Zaiee S, Moghadas Nejad F, Dareyni M, et al. Evaluation of rheological and mechanical properties of hot and warm mix asphalt mixtures containing electric arc furnace slag using gyratory compactor. *Constr Build Mater* 2023; 378: 131042.
[34] Choi YS, Lee SM. Progress in nuclear energy fundamental properties and radioactivity shielding performance of concrete recycled cathode ray tube waste glasses and electric arc furnace slag as aggregates. *Prog Nucl Energy* 2021; 133: 103649.
[35] Li S, Pan J, Zhu D, et al. A novel process to upgrade the copper slag by direct reduction-magnetic separation with the addition of Na 2 CO 3 and CaO. *Powder Technol* 2019; 347: 159–169.
[36] Zhang J, Qi Y, Yan D, et al. Characteristics and mechanism of reduction and smelting-separation process of copper slag. *J Iron Steel Res Int* 2015; 22: 121–127.
[37] Sarfo P, Das A, Wyss G, et al. Recovery of metal values from copper slag and reuse of residual secondary slag. *Waste Manag* 2017; 70: 272–281.
[38] Gyurov S, Marinkov N, Kostova Y, et al. Processing Technological scheme for copper slag processing. *Int J Miner Process* 2017; 158: 1–7.
[39] Luo S, Zhao S, Zhang P, et al. Co-disposal of MSWI fly ash and lead – zinc smelting slag through alkali-activation technology. *Constr Build Mater* 2022; 327: 127006.
[40] Xia Z, Zhang X, Huang X, et al. Hydrometallurgical stepwise recovery of copper and zinc from smelting slag of waste brass in ammonium chloride solution. *Hydrometallurgy* 2020; 197: 105475.
[41] Song S, Sun W, Wang L, et al. Recovery of cobalt and zinc from the leaching solution of zinc smelting slag. *J Environ Chem Eng* 2019; 7: 102777.
[42] Li Y, Qi X, Li G, et al. Removal of arsenic in acidic wastewater using Lead – Zinc smelting slag : From waste solid to As-stabilized mineral. *Chemosphere* 2022; 301: 134736.
[43] Ma Y, Li C, Yan J, et al. The release analysis of As and Cr metals in lead-zinc smelting slag : Mineralogical analysis, bioavailability and leachability analysis. *Environ Res* 2023; 229: 115751.
[44] Duan X, Li X, Li Y, et al. Separation and stabilization of arsenic in copper smelting wastewater by zinc slag. *J Clean Prod* 2021; 312: 127797.

[45] Hu H, Deng Q, Li C, et al. The recovery of Zn and Pb and the manufacture of lightweight bricks from zinc smelting slag and clay. *J Hazard Mater* 2014; 271: 220–227.
[46] Liu W, Du R, Zhao Z, et al. Development of lead smelting slag-based ecological mine backfill material : Performance control and heavy metals solidification. *Constr Build Mater* 2023; 367: 130310.
[47] Rashad AM, Mohamed H, Khalil MH, et al. Effect of micro-sized lead oxide on the workability, mechanical strength and durability of alkali-activated slag mortar. *Constr Build Mater* 2023; 373: 130890.
[48] Yadav Kanneboina Y, Saravanan T J, Ahmed Kabeer KIS, et al. Valorization of lead and zinc slags for the production of construction materials – A review for future research direction. *Constr Build Mater* 2023; 367: 130314.
[49] Li J, Liu Y, Ke X, et al. Geopolymer synthesized from electrolytic manganese residue and lead-zinc smelting slag: Compressive strength and heavy metal immobilization. *Cem Concr Compos* 2022; 134: 104806.
[50] Gholampour A, Danish A, Ozbakkaloglu T, et al. Mechanical and durability properties of natural fiber-reinforced geopolymers containing lead smelter slag and waste glass sand. *Constr Build Mater* 2022; 352: 129043.
[51] Sun Y, Hu S, Zhang P, et al. Microwave enhanced solidification/stabilization of lead slag with fly ash based geopolymer. *J Clean Prod* 2020; 272: 122957.
[52] Zhang X, Li L, Ul Q, et al. Preparation and characterization of glass ceramics synthesized from lead slag and lead-zinc tailings. *Ceram Int* 2023; 49: 16164–16173.
[53] Liu S, Cao X, Yang W, et al. Preparation of magnesium potassium phosphate cement from municipal solid waste incineration fly ash and lead slag co-blended: Ca-induced crystal reconstruction process and Pb-Cl synergistic solidification mechanism. *J Hazard Mater* 2023; 457: 131690.
[54] Gu M, Zhong Y, Wang L, et al. Separation of heavy metals from hazardous lead slag by carbothermic reduction and thermal volatilization–Effect of phase transformation on Sn, Pb, Zn removal. *Process Saf Environ Prot* 2021; 156: 330–339.
[55] Zhao Z, Liu W, Jiang Y, et al. Solidification of heavy metals in lead smelting slag and development of cementitious materials. *J Clean Prod* 2022; 359: 132134.
[56] Ünlü N, Drouet MG. Comparison of salt-free aluminum dross treatment processes. *Resour Conserv Recycl* 2002; 36: 61–72.
[57] Drouet MG, Handfield M, Meunier J, et al. Dross treatment in a rotary arc furnace with graphite electrodes. *Jom,* 1994, The Minerals, Metals & Materials Society (TMS) Proceedings; 46: 26–27.
[58] Drouet MG, Leroy RL, Tsantrizos PG. *Drosrite salt-free processing of hot aluminum dross*. Pittsburgh, Pennsylvania, 2000, pp. 1135–1145.
[59] Meshram A, Singh KK. Recovery of valuable products from hazardous aluminum dross: A review. *Resour Conserv Recycl* 2018; 130: 95–108.
[60] Meshram A, Singh KK. Generation of hydrogen-gas from aluminum dross. In: *European Metallurgical Conference 2017*. Leipzig, Germany, 2017, European Metallurgical Conference Proceedings, pp. 1451–1460.
[61] Meshram A, Jain A, Gautam D, et al. Synthesis and characterization of tamarugite from aluminium dross: Part I. *J Environ Manage* 2019; 232: 978–984.
[62] Meshram A, Jain A, Rao MD, et al. From industrial waste to valuable products: Preparation of hydrogen gas and alumina from aluminium dross. *J Mater Cycles Waste Manag* 2019; 0: 0.
[63] Singh KK, Meshram A, Gautam D, et al. Hydrogen production using waste aluminium dross : From industrial waste to next-generation fuel. *Agron Res* 2019; 17. DOI: 10.15159/AR.19.022Hydrogen

[64] Ma N, Houser JB. Recycling of steelmaking slag fines by weak magnetic separation coupled with selective particle size screening. *J Clean Prod* 2014; 82: 221–231.
[65] Owais M, Yazdani MR, Mika J. Detailed performance analysis of the wet extractive grinding process for higher calcium yields from steelmaking slags. *Chemical Engineering and Processing - Process Intensification* 2021; 166: 108489. DOI: 10.1016/j.cep.2021.108489
[66] Owais M, Yazdani RM, Järvinen M. Wet extractive grinding process for efficient calcium recovery from steelmaking slags. *Chem Eng Process Process Intensif* 2020; 151: 107917.
[67] He Z, Hu X, Chou K. Oxidative modification of industrial basic oxygen furnace slag for recover iron-containing phase: Study on phase transformation and mineral structure evolution. *Process Saf Environ Prot* 2023; 171: 167–175.
[68] Diao J, Zhou W, Ke Z, et al. System assessment of recycling of steel slag in converter steelmaking. *J Clean Prod* 2016; 125: 159–167.
[69] Teo P, Abu A, Basu P, et al. Recycling of Malaysia's electric arc furnace (EAF) slag waste into heavy-duty green ceramic tile. *Waste Manag* 2014; 34: 2697–2708.
[70] Lee Y, Kim H, Kim M, et al. Bond performance of reinforced concrete beams with electric arc furnace slag aggregates. *Constr Build Mater* 2020; 244: 118366.
[71] Lopez Ferber N, Al Naimi KM, Hoffmann J, et al. Development of an electric arc furnace steel slag-based ceramic material for high temperature thermal energy storage applications. *J Energy Storage* 2022; 51: 104408.
[72] Stewart DJC, Barron AR. Resources, conservation & recycling pyrometallurgical removal of zinc from basic oxygen steelmaking dust – A review of best available technology. *Resour Conserv Recycl* 2020; 157: 104746.
[73] Sukmak P, Sukmak G, Silva P De, et al. The potential of industrial waste : Electric arc furnace slag (EAF) as recycled road construction materials. *Constr Build Mater* 2023; 368: 130393.
[74] Kim J, Azimi G. Valorization of electric arc furnace slag via carbothermic reduction followed by acid baking–Water leaching. *Resour Conserv Recycl* 2021; 173: 105710.
[75] Zhou H, Liu G, Zhang L, et al. Mineralogical and morphological factors affecting the separation of copper and arsenic in flash copper smelting slag flotation beneficiation process. *J Hazard Mater* 2021; 401. DOI: 10.1016/j.jhazmat.2020.123293
[76] Xu D, Fu R, Tong Y, et al. The potential environmental risk implications of heavy metals based on their geochemical and mineralogical characteristics in the size-segregated zinc smelting slags. *J Clean Prod* 2021; 315: 128199.
[77] Xu D, Fu R. The mechanistic understanding of potential bioaccessibility of toxic heavy metals in the indigenous zinc smelting slags with multidisciplinary characterization. *J Hazard Mater* 2022; 425: 127864.
[78] Lin H, Weng W, Zhong S, et al. Enhanced recovery of zinc and lead by slag composition optimization in rotary kiln. *Trans Nonferrous Met Soc China* 2022; 32: 3110–3122.
[79] Rachwał M, Magiera T, Wawer M. Coke industry and steel metallurgy as the source of soil contamination by technogenic magnetic particles, heavy metals and polycyclic aromatic hydrocarbons. *Chemosphere* 2015; 138: 863–873.
[80] Wang J, Huang Y, Li T, et al. Contamination, morphological status and sources of atmospheric dust in different land-using areas of a steel-industry city, China. *Atmos Pollut Res* 2020; 11: 283–289.
[81] Kicinska A, Wikar J. Ecological risk associated with agricultural production in soils contaminated by the activities of the metal ore mining and processing

industry–Example from southern Poland. *Soil Tillage Res* 2021; 205, 104817. DOI: 10.1016/j.still.2020.104817

[82] Nahar Jannat J, Mia YM, Mastura Munia Farajana Jion M, et al. Pollution trends and ecological risks of heavy metal (loid)s in coastal zones of Bangladesh : A chemometric review. *Mar Pollut Bull 2023*; 191, 114960.

[83] Izydorczyk G, Mikula K, Skrzypczak D, et al. Potential environmental pollution from copper metallurgy and methods of management. *Environ Res* 2021; 197: 111050.

[84] Ji R, Liu T, Kang L, et al. A review of metallurgical slag for efficient wastewater treatment : Pretreatment, performance and mechanism. *J Clean Prod* 2022; 380: 135076.

[85] Mele M, Magazzino C. A machine learning analysis of the relationship among iron and steel industries, air pollution, and economic growth in China. *J Clean Prod* 2020; 277: 123293.

[86] Fukuzawa R. Climate change policy to foster pollution prevention and sustainable industrial practices – A case study of the global nickel industry. *Miner Eng* 2012; 39: 196–205.

[87] Di Maria A, Merchán M, Marchand M, et al. Evaluating energy and resource efficiency for recovery of metallurgical residues using environmental and economic analysis. *Journal of Cleaner Production* 2022, 356, 131790.

[88] Kumar Acharya P, Kumar Patro S. Evaluation of environmental disturbance indicator using functional performance and life cycle assessment of ferrochrome waste concrete. *J Build Eng* 2023; 65: 105788.

[89] Mankhand TR. Recovery of valuable materials from aluminum dross. *J Sustain Planet* 2012; 3: 86–94.

[90] Izadi A, Mohebbi A, Amiri M, et al. Removal of iron ions from industrial copper raffinate and electrowinning electrolyte solutions by chemical precipitation and ion exchange. *Miner Eng* 2017; 113: 23–35.

[91] Rudnik E. Hydrometallurgical recovery of zinc from industrial hot dipping top ash. *Trans Nonferrous Met Soc China* 2020; 30: 2239–2255.

[92] Rudnik E. Recovery of zinc from zinc ash by leaching in sulphuric acid and electrowinning. *Hydrometallurgy* 2019; 188: 256–263.

[93] Singh S, Ramakrishna S, Gupta MK. Towards zero waste manufacturing : A multidisciplinary review. *J Clean Prod* 2017; 168: 1230–1243.

[94] Farzadkia M, Jorfi S, Nikzad M, et al. Evaluation of industrial wastes management practices : Case study of the Savojbolagh industrial zone, Iran. *Waste Manag Res* 2020; 38: 44–58.

[95] Srivastava RR, Rajak DK, Ilyas S, et al. Challenges, regulations, and case studies on sustainable management of industrial waste. *Minerals 2023*; 13, 51.

5 Metal Extraction and Beneficiation Wastes

Arunabh Meshram

5.1 INTRODUCTION

A world without metals to facilitate mankind is unimaginable today. Metals have played a crucial role in shaping the development of technologies of the modern world. It is, therefore, important to understand the extraction and application of metals. The conventional methods of metal extraction and application are widely described in literature; this chapter involves the recovery of metals, alloys, and materials from metal-rich wastes.

A complete overview of metallurgical waste management is essential for understanding the intricate subtleties of metal recovery and metal cycle. In Chapter 4, the recovery of valuable materials from metallurgical wastes has been discussed. The present chapter broadens the perspective of materials recycling in metallurgical industries. To comprehend the methods of metal extraction from metallurgical wastes, important processes will be discussed, along with further investigation of types of wastes. Metal waste reclamation and its reintroduction into the material cycle is the primary theme of this chapter.

As discussed previously, there are three major classes in metallurgical industries. With pyrometallurgical, hydrometallurgical and electrometallurgical processes, the generation of waste is inevitable. Figure 5.1 discusses the general overview of any metallurgical industry involved in metal extraction and production. The mineral ore beneficiation done at the initial stage of mining and partly at the industry is one of the most important stages of mineral concentration. The separation of gangue and other lesser important minerals from the ore is achieved in this stage.

In many of these processes, water resources are used for wet comminution like grinding etc. These processes are used to improve metal-rich mineral separation. These slurries should also be properly treated after the extraction of valuable minerals. Mining wastes have also been discussed in the upcoming sections to understand the economic importance of these wastes. Usually, these wastes are landfilled or released into water streams. Such wastes, just like the wastes generated in industries, are rich in valuable materials that can be extracted by suitably designed processes.

The second chief step of the metal cycle involves the primary extraction and production stage. The stage involves many inherent steps like ore pretreatment (roasting, calcination, etc.), extraction (pyro-, hydro- or electrometallurgical route) and refining

DOI: 10.1201/9781003619680-5

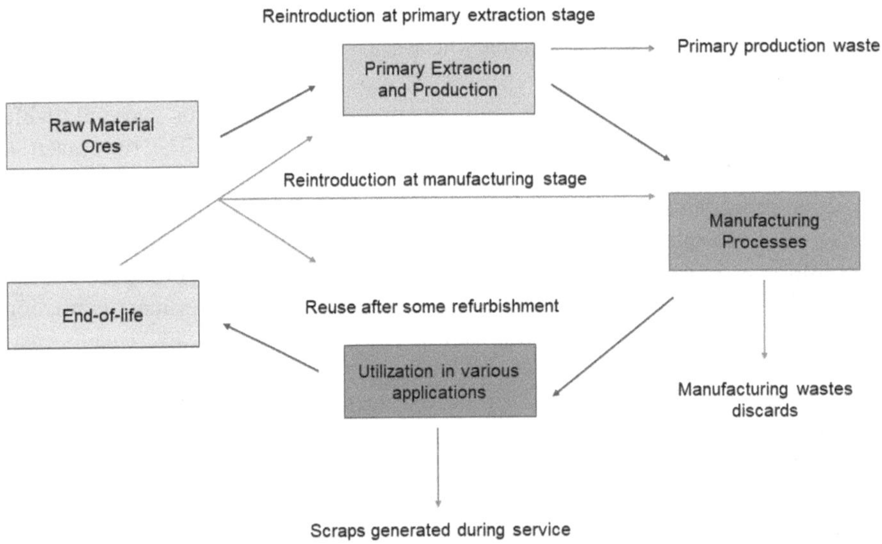

FIGURE 5.1 A generalized schematic diagram of material flow from its raw materials to its end-of-life.

(fire, electrorefining etc.). The generation of waste in this stage has been discussed in the previous chapter. The resultant metal is transferred to the next stage of manufacturing. These involve oriented manufacturing processes to prepare the extracted metal ready for their desired applications. These processes involve machining, rolling, forging, welding, extrusion, wire drawing etc. It is also possible that at this stage, a secondary metal extraction/ refining step is involved. The reutilization of metal wastes from various sources can be done by remelting and casting of metal-rich products. In these manufacturing processes, production rejects are generated. The finished metal products are transported and used in their intended applications.

The next important stage in the metal cycle is in the application. As has been discussed, construction, structural, aviation, strategic and automation applications of metals are well documented in literature. With metal products serving mankind in these areas the development of the present-day world has escalated manifold. After the intended application, the metal product is discarded or removed from the service, and it reaches the end-of-life stage.

The most crucial challenge in any metal industry is to ensure sustainable production with economic benefits, while reducing the environmental burden and dependence on natural resources. To facilitate the continuity of metal production, a constant supply of raw materials is essential. This reduces the possibility of raw material crisis at the production end. As procuring natural ores from mining sites requires investment in logistics and pretreatment, many alternative sources of metal have been explored.

It is important to note that many non-ferrous metals, especially aluminum, have been recycled numerous times. Even with the recycling of metals, the general demand for these metals in various applications has increased with passing years. In the upcoming sections, the classification of metal discards and wastes will be discussed. Placing proper economic value on metallic wastes has inclined the development of modern processes.

As discussed in Figure 5.1, the reintroduction of metal discards and end-of-life products into the metal cycle at various stages is very crucial. The raw material balance at the production end can be achieved by using these metal discards. Primary extraction processes can include the metal waste after appropriate refining, as these wastes can have other impurities like unwanted trace elements. Obtaining proper composition is necessary for further processing of metal. The alloying operation can become difficult, if the composition of metal produced during the primary extraction process varies due to introduction of metal wastes and scraps.

Similarly, the reintroduction at the manufacturing stage will necessitate the remelting and refining using appropriate pyrometallurgical facilities. Making the composition of the finished product as per the demands will be necessary. In both of the inclusions of metal wastes, the utilization of metallic value is possible. A simpler method of reusing end-of-life wastes is refurbishing the finished product; however metal recycling is achieved with greater efficiency in the first two routes, i.e., at the primary extraction stage and manufacturing stage.

This chapter connects the utilization of metal after it reaches its end-of-life. Some important examples of metallurgical wastes that are of economic and environmental importance will be discussed. Environmental concern of waste management will be highlighted. Development of advanced recycling routes, involving multiple stages of unit processes will be described, along with the application of these routes for successful metal recycling. Some discussion on implementing modern technologies to conventional processes has been included for developing insight into the challenges faced by the production industries. Modern waste management system ensures efficient metal recovery; the final section is dedicated to waste management systems.

5.2 TYPES AND SOURCES OF METAL WASTE

As has been described in the previous chapters, wastes that are discarded by the human population can be characterized to ascertain the economic value of the inherent recyclable materials. One of the key aspects of materials recycling involves the economic outcome of the process. If the economic benefits of recycling a given waste are not compensating for the operation cost, the recycling route becomes infeasible on a larger scale of operation. Therefore, engineering economics plays a crucial role in material recycling. A similar trend is observed for metal industries and metallurgical wastes.

It is essential to understand the importance of mineral beneficiation to appreciate metal production and its cycle. Minerals are the source of valuable metals and developing an efficient method of utilizing the minerals for metal extraction is a part of the modern process. Due to mining activities directed to provide for the ever-increasing demand for raw materials for various industries, the subsurface of the Earth is widely affected. Using the mining wastes, these altered substructures can also be repaired [1].

While it is important to examine the metal production routes, it is equally important to explore the important methods of mineral processing technologies that are used for metal recovery from mining discards. Whitworth et al. have detailed the key trends on the processes used for the metal recovery from these wastes [2]. Crushing, milling, classification processes (flotation and gravity separation) are the unit operations done before the ores are used for metal extraction in metallurgical industries. Newer methods for mineral beneficiation include stirred mills, semi-inverted mills, semi-inverted hydrocyclones, fluidized beds with multiple reactors and compartments for facilitating the separation process and counter-current fluidized bed separation process.

Wastewater treatment is an important aspect that is associated with many metallurgical and allied industries, like coal treatment plants. Controlling pollution by wastewater treatment is being explored by many researchers. Shi et al. have investigated the key challenges in the wastewater treatment originating from coal industries [3]. According to their findings, the implementation of local laws of environmental protection should be properly supervised, sufficient subsidy should be provided for the treatment of highly saline wastewater. The implementation of environmental protective measures should be done at the industrial scale and included in the management of the coal plants; extensive research should be encouraged for environmental safeguarding. Cohesive efforts from various stakeholders and the general public should be encouraged for better recycling results.

From the yesteryears, the utilization of mining and beneficiation wastes in the construction applications has been a method of material consumption. The environmental hazards associated to the landfilling/ open dumping of these beneficiation wastes are avoided when used as inert material for construction. While some of these wastes are used for making concrete, others are fit for bricks and tiles etc. Inert applications ensure material utilization with effective output [4]. Following the advances, the beneficiation and mining waste are also used for a wider range of construction materials [5]. Tailings and waste rocks are combined to use as raw material for cementitious applications, whereas the slags from various metallurgical industries are used for making fired bricks.

Mining waste rocks have been used for the production of fly-ash based geopolymers [6]. Using the alkaline precursor solution, the homogenized mixture of raw materials was blended with binders, molded, and cured. It has been reported that with the mining waste, the desired strength of the geopolymer was attained and the aggregate (mining waste) contributed to it. Similarly, lead-bearing crystals and steel slag have been used for the preparation of alkali activated cement [7]. The raw materials were ball milled and mixed with alkaline solution of NaOH, molded and cured. This method is another alternative to the conventional disposal of beneficiation waste.

Taking an example of global aluminum production, it has been reported that in 2021, around 37 million tons of aluminum was accounted for recycled production. Some future projections indicate that by 2070, 120 million tons of recycled aluminum alloy production will be achieved. The variation of scraps available for aluminum recycling is also interesting; 22.2 million tons of old scrap was available in 2021, compared to 15.5 million tons of new scrap. This indicates that aiming old scrap for metal extraction can lead to higher metal recovery [8].

An advantage evident for metal industries is the recyclability of metals. There are well-established methods of metal refining implemented at the industrial scale. Fire refining and electrorefining are commonly used refining processes in many non-ferrous industries. The reutilization of metallic wastes as raw materials for metal extraction/refining has been explored by many researchers. Availability of enormous metallic wastes for recycling has initiated these methods.

The characterization of scraps and metal discards is necessary to determine the extent of refining necessary to achieve the final composition. Scraps can be categorized on the basis of their origin as in-house scraps, in-service scraps and end-of-life scraps. In-house scraps are produced in the metal extraction and manufacturing stages. Essentially, to achieve higher metal production, the utilization of in-house scraps is highly encouraged in these stages. In-service scraps are generated at the location of service, due to the damage taken during the application of the product. Some defects can be attributed to improper transportation, whereas others can be generated due to various service conditions, for example, corrosive environment. End-of-life scraps are produced after the intended application of the product is completed or the product reaches the completion of its service life as per the safety standards.

Figure 5.2 highlights the key classification of scraps based on various parameters. As has been noted, ferrous scraps include the scraps containing iron and steel as the primary metal, whereas non-ferrous scraps can have aluminum, copper, magnesium, zinc, and other metals as the chief metal. The scraps can be coated with paint and other protective materials. Such scraps will require additional pretreatment to be fit for further processing.

Scraps can also be categorized on the basis of origin as new or old scrap. The scraps produced in an industry during primary or secondary extraction or manufacturing stage are called new scraps. These are readily available for metal recovery. Most of the industries are inclined to using these scraps as secondary sources of metal. The scrap generated after a given finished product has reached its end-of-life (after service) is called old scrap. Also, these old scraps can be collected, stored, sold, or landfilled, making their reclamation a challenge for the recyclers. The depletion of

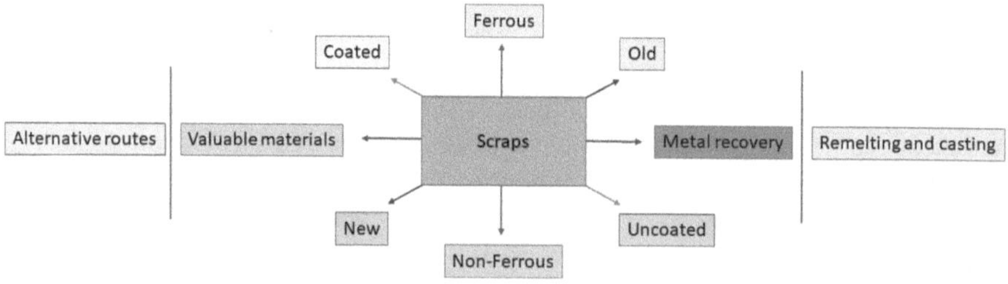

FIGURE 5.2 Classification of scraps based on type of metallic values present, age, and physical characteristics.

metallic values from these old scraps due to environmental conditions is evident and such scraps require further pretreatment to make them fit for reutilization.

To ensure higher metal recycling and recovery, scrap managers and metal industries should collaborate. As discussed in the upcoming sections, the industries are highly dependent on the availability of high-quality scraps. The presence of impurity alloying elements in the metal is a drawback for processing the low-quality scraps. The separation of these alloying elements from the scraps is very difficult.

The method of scrap separation is a trial for any scrap manager, whether independent or associated with an industrial partner. In many countries, scrap management is not well organized, making the scrap collection and sorting an arduous task. Many technological advances have been observed in recent years. Computational advances, deep learning and simulation have been incorporated into the sorting of scraps. The reduction manual labor and human exposure to the hazardous scraps is being aimed at, with emphasis on the coupling multiple methods of pretreatment to recycle the scrap stream. Some of the important scrap processing steps involved are magnetic separation, heavy media separation, eddy current separation along with visual inspection using advanced imaging techniques.

Figure 5.2 also indicates the method of recycling these scraps. One of the chief methods is the reclamation of metallic values from the scraps by remelting. There are many other methods being explored by researchers. These methods aim for the generation of valuable products from these scraps. By employing some unit processes from pyrometallurgical, hydrometallurgical and electrometallurgical routes, valuable materials can be produced. Some of these methods are described in the upcoming sections.

Although metallic values in scraps are significantly very large and these are directly used in many industries, the presence of metallic values in other wastes like dross, skims, and other industrial wastes broadens the focus of material recovery. Some of these metallurgical wastes have been discussed in Chapter 4. In the next section, the scraps and other metallic wastes have been described.

5.3 ACCOUNTING METAL WASTES AS RESOURCES

From the previous section, an overview of metallic values present in various metal industry wastes is developed. A highlight of the scraps has already been presented. Through tables, the scraps and their recycling processes will be discussed in this section. Some of these processes like remelting and casting are commonly applied in the industries as the facilities for these processes are readily available. However, newer processes with higher metal recovery have also been reported.

Tables 5.1 to 5.4 describe various scrap materials generated around the metal industry. It is important to note the metallic concentration in these scraps. When these scraps are taken as secondary sources of metals, the efficient extraction of metallic values leads to recirculation of these metals. In this manner, the scraps are reintroduced as resources and not as waste. A wide variety of products are also generated when these scraps are subjected to different recycling processes.

TABLE 5.1
Aluminum waste and its applications

Type of metal scrap	Method of recycling	Recycled product	References
Aluminum scrap	Scraps were washed, briquetted, dried, and melted in electric resistance furnace	Aluminum ingots	[58]
	Friction stir consolidation process (vertical die loading coupled with rotation) for sheet metal scraps	Billets of scrap after friction stir process	[59]
	Centrifugation and purification of aluminum alloy scrap by super-gravity technology and semi-solid processing (6061 and 1060 alloys)	Pure aluminum and separation of impurities like Fe, Cu, Mn, Mg, Si etc.	[60, 61]
	Cold compaction of aluminum scrap particles generated after machining operations	Aluminum powder Compacts	[62]
	Shear assisted Processing and Extrusion (ShAPE) method used for tolerating Fe content in waste aluminum scrap. Pothole die extrusion also used for 6063 aluminum alloy	Aluminum billets	[63, 64]

According to the findings of Van den Eynde et al., the surplus of aluminum scrap at the global scale will reach ~8.7 million tons by 2040 [9]. This estimate has been done by incorporating various parameters like determining the alloy composition of the scraps into the International Aluminium Institute model.

Similarly, Raabe et al. have detailed the utilization of scraps by describing the insights in waste management [10]. With the wide variety of aluminum alloys that form the composition of the scraps, sorting becomes very crucial. Important stages like comminution, heavy media separation and magnetic separation are key steps for scrap sorting, along with eddy current separation. These are also followed by packaging and briquetting stages.

Sevigné-Itoiz et al. explored the material flow of industrial aluminum scraps from European countries to other countries [11]. By consequential life cycle assessment, it is observed that greenhouse gases can be regulated if the old scraps are recycled locally, instead of exporting. Similarly, Hatayama et al. analyzed aluminum scrap flow

TABLE 5.2
Magnesium waste and its application

Type of metal scrap	Method of recycling	Recycled product	References
Magnesium scrap	Primary alloy ingots and diecast magnesium scraps (AZ91D) were used for rheo-diecasting process	Diecast magnesium	[65]
	High pressure die casting was used to prepare various alloys using AM50 alloy	Magnesium secondary alloy	[66]
	Casting of magnesium alloy using primary alloy (AZ91), varying the magnesium scraps content	Cast magnesium alloy	[67]
	Aqueous NaCl catalysis with Pt-coated Ti and stainless steel net was done for low-grade magnesium scrap to produce H_2. Organic acids were also used	H_2 evolution	[68, 69]
	Melt conditioned high pressure die casting process used for recycling AM50A and AM60B magnesium scrap	Cast Magnesium alloy	[70]
	AZ91D flash scrap used for Ni-plating from wastewater by electrolysis	Ni^{2+} concentration reduced in wastewater	[71]
	Magnesium chips leached with HCl and precipitated with NaOH	Magnesium hydroxide aerogels	[72]
	AZ91 Magnesium chips remelted to achieve higher refining by fusion with fluxes	Cast Magnesium with reduced impurity	[73]
	AZ91D magnesium scraps subjected to cold pressing, heat treatment and hot extrusion	Extruded Mg Alloys	[74–76]
	AZ91D magnesium scrap used for vacuum gasification and distillation	High purity magnesium	[77]
	Magnesium black dross from AE44 alloy crushed and downsized, water leached, acid leached and subjected to REE recovery	REE like Ce, La, Nd, Pr recovered	[78]

to understand the potential of improving recycling of metallic scraps [12]. Reduction of CO_2 emissions is necessary for reducing environmental burden.

Zhu et al. have investigated the generation of aluminum scraps from aluminum sheet incentive automobiles in the United States [13]. It is reported that proper sorting of the automobile scraps is necessary to achieve higher metal recovery. It is projected that by 2050, 246 kt/year scrap will be generated.

TABLE 5.3
Copper scrap and its recycling

Type of metal scrap	Method of recycling	Recycled product	References
Copper scrap	Scraps of copper melted with other scraps to produce Cu-Pd alloy followed by recovery by electrodeposition	Palladium recovery	[79]
	Sulfurization-vacuum distillation process used for removing impurities from copper scrap	Pb, Zn and Sn separated from scrap copper	[80]
	Copper recovery from E-waste by bioleaching	Copper recovery	[81]
	Copper scrap used as raw material for copper powder production by galvanostatic electrolysis	Fine copper powder	[82]
	Copper scrap anode slime was roasted, leached and precious metals were recovered	Cu, Ni, As were recovered	[83]

Similar to aluminum, the recycling of magnesium scraps and waste is described in Table 5.2. Being a metal of immense importance due to its relatively lower density, magnesium alloys, along with aluminum alloys, are crucial for modern applications. The key sources of scrap are the remelting of alloy and manufacturing unit that generate machine operation discards. It is important to efficiently recycle these scraps. In some modern processes, complete recycling of metallic magnesium is aimed, for instance, applying vacuum distillation process for separation of impurity elements.

The energy consumption involved in recycling also plays an important role while developing a new process for recycling. It is reported that recycling of scraps only consumes 5% of total energy required for primary magnesium production. This indicates that recycling of metallic values is an economically viable route to meet the growing demands of the world. Magnesium alloys are being used in automotive industries. A sheer rise in these sectors has led to an increase in the demand for these alloys.

Copper scraps are usually generated at various sources, including manufacturing shops, casting units and alloying furnaces. Table 5.3 represents a holistic recycling outline for these scraps. A wide range of methods have been described, including recovery of impurity elements from the scraps stream to recover copper and generation of fine copper powder by galvanostatic electrolysis.

A similar insight into the recycling of zinc waste is presented in Table 5.4. A good source of zinc is the BOF dust produced in the steelmaking process. Efficient extraction of zinc from this dust is necessary. The application of zinc is primarily in the galvanization of various steel products. The presence of zinc in the BOF steelmaking dust is essentially due to the recycling of galvanized steel scraps and wastes. It has

TABLE 5.4
Zinc wastes and its recycling

Type of metal waste	Method of recycling	Recycled product	References
Zinc wastes	Zinc dust from steelmaking industries subjected to carbothermic reduction	Fe and Zn	[14, 84]
	Zinc dust and iron oxide milled, briquetted, roasted and leached with HCl	Zinc ferrite	[85]
	Zinc oxide dust leached with sulfuric acid and iron powder under ultrasonication	Germanium recovery	[86]
Zinc ash	Zinc ash leached with H_2SO_4	Zinc recovery	[87, 88]
	Zinc ash, zinc oxide and white flour used as admixtures for cement and mortar	Concrete	[89]
Zinc scrap	Zinc scrap used for preparation of ZnO nanostructure by controlled oxidation	ZnO nanostructures	[90]
Zinc dross	Zinc dross leached with sulfuric acid followed by addition of $(NH_4)_2HPO_4$ to produce zinc-rich materials	$Zn_3(PO_4)_2.4H_2O$ and $(NH_4)_2SO_4$	[91, 92]
	Zinc dross melted, cast, and machined for super-gravity separation	Zinc recovery	[93]
	Zinc dross used as raw material for anode for electrolysis to recover zinc	Zinc powder	[94]
	Zinc dross melted and cast as anode for electrorefining in NH_4Cl-NH_3 electrolyte	Refined zinc	[95]
Zinc solder dross	Leaching of zinc solder dross, followed by precipitation of various valuable materials from leach liquor and solid residue	$ZnSO_4$, $ZnCO_3$, SnO_2. xH_2O and lead hydroxide	[96]

been observed that EAF dust has higher zinc concentrations than the BOF dust, as scrap intake for EAF is higher than the BOF [14].

The categories of zinc dust based on the zinc concentration are low-zinc (Zn <8 wt.%), medium-zinc (8 wt.% < Zn < 20 wt.%) and high-zinc (Zn > 20 wt.%). The disposal of zinc dust can be done by pyrometallurgical and hydrometallurgical routes, apart from landfilling it. Given the rising concern on environmental damages due to landfilling of metallic wastes, zinc dust is recycled by pyrometallurgical route, as it leads to metal recovery and economic gains.

The key operation of metal recovery from the EAF dust involves carbothermic reduction, leading to the conversion of metal oxides to metals. Selective reduction roasting has also been explored for the recovery of zinc from the dust [15]. The reduction of $ZnFe_2O_4$ is essential for hydrometallurgical operations. Pretreatment operation involving the combination of reduction roasting and water leaching has been reported to be beneficial for high-zinc EAF dust.

Similar to zinc scraps and EAF dust, zinc dross has been used as a secondary source of metal. As a method of removing chloride from the dross, a pretreatment involving washing has been investigated by Guresin and Topkaya [16]. The influence of washing dross with distilled water is prominent; nearly 80% chlorine present in the zinc dross was reported to be removed.

With the natural resources used for the production of metal in extraction stage, it is also important to consider alternative resources for supplementing the production. Although the question of direct utilization of such resources always remains, the possibility of efficient metal recovery is highly evident, given the availability of a variety of scraps and other metallic wastes in and around these industries. Also, many industries have some key unit processes that facilitate metal extraction. These processes are usually pyrometallurgical in nature, leading to high-temperature operations. To scale up alternative processes from hydrometallurgy and electrometallurgy, the efficiency of metal extraction needs to be ensured.

As observed above, a large variety of metal scraps and other metal sources are present for the recovery of metals. Designing an efficient method is of utmost importance. Such processes need to be compatible with the type of scraps and other metal sources available in abundance and the technology driving these processes should be economical. Waste metal recycling processes should be adding value to the overall metal cycle, rather than being a source of additional waste that requires further processing. Process economics should be studied thoroughly to apply the optimization for higher metal recovery with given resources.

5.4 INDUSTRIAL METAL WASTE PROCESSING TECHNOLOGIES

As discussed in the previous sections, the scraps and other metallic wastes have been transformed into secondary resources for efficient metal recovery. However, implementing a dedicated recycling process for metal recovery from such wastes to achieve sustainability is a hurdle that modern industries must overcome in the coming years. In this section, some of the modern processes are highlighted that can have an impact on metal recovery at industrial scale.

There have been investigations of steel scraps to understand the composition and applications. One such study has described the presence of copper precipitates in low carbon steel rebars produced after recycling steel scraps [17]. These copper precipitates are brittle and lead to a reduction in the mechanical properties of the steel. By addition of Si and Mn to the steel can change the precipitates' composition and overall steel properties. The most applicable method for scrap utilization in the steel industry is the remelting in the Electric Arc Furnace (EAF). It leads to maximum recovery of metallic content present in the scrap, keeping the loss only limited to the slag phase.

The method of reintroduction of scraps into the steel cycle is well documented in literature and practiced across the steel industries around the globe. The optimization of scrap remelting usually requires the consideration of scrap weight, category, and presence of impurities. The chemical composition of the resultant steel needs to be attained before it is ready for the next stages of refining. This can be done with the addition of molten pig iron to the EAF. The optimization of scrap quantity is an important step for the achievement of the desired final composition, as many impurities are associated with these scraps.

Using EAF steelmaking unit, the production of alloy steels has escalated immensely. One of the key factors contributing to the alloy steelmaking is the utilization of scraps (new, in-house scraps and old, end-of-life scraps). With the addition of these scraps during alloy steelmaking, lesser quantities of ferroalloys are consumed for maintaining the quality of the product. This leads to efficient use of resources and reduction in the cost of production, as in-house scraps are easily available for remelting.

Apart from the remelting of scraps in the EAF, some other applications of steel scraps have been reported recently. These are presented in Table 5.5. As can be seen, steel scraps are used for a variety of applications, including recovery of hazardous materials from wastewater streams and production of high-strength concretes.

To investigate the characteristics of fluoride salts as fluxing material for aluminum scrap melting, Wan et al. have used KF, K_3AlF_6, $KAlF_4$ and AlF_3 as potential raw materials [18]. It has been found that KF, K_3AlF_6 and $KAlF_4$ have relatively better oxide removing ability, compared to AlF_3. This suggests that presence of fluoride in the flux can lead to better oxide layer removal from the scrap metal during melting. On the other hand, Xiao et al. used NaCl-KCl-Na_3AlF_6 as the salt flux for melting scraps of various compositions [19]. According to their findings, the metal recovery from molten scrap is higher when the concentration of fluorides is higher.

Scraps generated from the end-of-life aircraft have been used for the remelting, followed by composition readjustment and casting [20]. These remelted and cast

TABLE 5.5
Steel scrap recycling and modern advances in recovery

Type of metal scrap	Method of recycling	Recycled product	References
Steel scrap	Sulfurization-vacuum decomposition of steel wire scraps using pyrite	High purity tin sulfide from steel scraps	[97]
	Steel scrap used as reinforcement material for concrete	Concretes with higher strength	[98]
	Slags generated from steel scrap recycling used for the recovery of arsenic (V), Cd and Pb from contaminated water under various pH values	Recovery of arsenate from wastewater	[99, 100]

alloys have been further extruded and heat treated to achieve the desired physical properties. AA7075 alloys have been reported to be produced by this method. Similarly, Mashhadi et al. have subjected AA336 turning scraps to cold sintering [21]. The compacts were subjected to melting with salt flux additions and it was found the resultant cast product had desired physical and mechanical properties.

Gomez et al. have described a model for estimating copper scraps and its economic value [22]. The secondary method of production is also utilizing old scraps. The flow of scraps to these processes can be regulated to assure higher metal recovery.

To recycle magnesium scraps with coatings, it is important to de-coat and clean the raw material before using it for casting applications. Exergy analysis of coated magnesium scrap recycling has been done by Meskers et al. [23]. It is observed that the exergy accounts for chemical composition changes, oxidation losses, alloying and generation of intermetallic compounds. Categorization of coated magnesium alloys can further help in improving the recycling efficiency, as it can help in sorting these scraps better.

The EAF steelmaking has been explored by many researchers across the world to achieve higher metal recovery. A deeper insight into the area of automotive industry for the utilization of scraps has been presented by Diener et al. [24]. Material flow analysis (MFA) for the automotive industry presents a challenge, as the sorting of the scraps for direct utilization and component remaking becomes a key step. The sorted components could be reused for specific applications.

Achieving a circular economy is not limited to a single industry or a sector of industries. It is a goal that needs to be achieved collectively by human society. A concerted effort towards this goal can lead to better metal recycling in the manufacturing sector. The collaboration of various industries like steelmaking industries and scrap suppliers is one such example. Berlin et al. have tried to explain this connection for Sweden [25]. To achieve a circular economy at a larger scale, continued supply of scraps to the developed steelmaking industries in Sweden is important, as these industries rely on the supply of scraps.

5.5 CONVENTIONAL TECHNOLOGICAL CHALLENGES

In the present section, the key challenges to waste recycling of metal-bearing materials have been highlighted. These challenges indicate the potential for further improvements in existing technologies and result in modern advances in conventional processes.

It is important to understand that the resources present on the Earth are not infinite and material recycling is necessary to sustain the processes. According to Wang et al., copper resources are being regulated such that the recycling of copper increases to recover copper from ever-increasing copper scraps [26]. Developing processes focused on recycling is one of the key challenges presently. Similarly, Xi Tian et al. and Di Dong et al. have described the impact of regulation of copper resources on interregional logistics of copper scraps [27, 28]. The concentration of copper scraps in the coastal areas of China will be distributed throughout the country. Shuo Liu et al. and Wang Junbo et al. have indicated the gap between the production and consumption of copper in China and emphasized on the recycling of copper scraps [29, 30]. Such scraps are essentially secondary resources for the extraction of copper.

In North America and Europe, the material flow is playing an important role in copper recycling [31]. The characterization of scraps is necessary for acceptance at the industrial scale for efficient metal refining. The production of new scrap is also decreasing, as the primary production industries are becoming increasingly efficient. Scrap sorting for copper-rich scrap is still difficult. To analyze the material flow of copper scraps, Wang et al. and Hu et al. have considered the flow of copper scraps from developed to developing countries [32, 33]. It has been observed that India can emerge as a potential market, as it is one of the key emerging economies of the world.

With the rise in complexity of modern alloys dedicated to specific applications, the recycling of copper alloy scraps becomes very difficult [34, 35]. Attaining an optimum metal recovery from waste scraps requires additional processes for each alloying element, adding to the cost of recycling. Reducing the production and consumption of copper can lead to reduction of GHG emissions, however the focus should be on the end-of-life copper-rich waste recycling. Conventional copper production generates higher pollutants compared to environmentally friendly recycling processes [36]. The energy consumption of ore mining and its processing is significant.

Another important challenge to modern recycling processes is the nature of metal being recycled [37]. For instance, Mg and Al are alkali metals. These metals have an inherent tendency to react with the surrounding environment under certain conditions. For example, aluminum dross/ scraps with fresh aluminum surface can react with alkaline or acidic solutions. This can lead to the generation of many gases like H_2, CH_4, and NH_3, making the recycling process more complex [38, 39]. These gases are hazardous and proper contingency is necessary for safe disposal of these gases.

When recycling of Mg automotive scraps is considered, the environmental footprint is essentially related to the emissions of CH_4, CO, NO_x, and SF_6 [40]. These emissions are hazardous to the environment; a projected recycling emission of ~5.6 MT of CO_2 eq. is predicted by 2050.

While considering steel scrap recycling, one of the most important aspects that impacts the continuity of the process is the availability of good quality scraps. To address this, Liu et al. have presented an evolutionary game model for the simulation of the quality scrap supply from various sources and its utilization in the steel sector [41]. The inclusive contribution of steel sector and scrap steel sectors is very important for the development of advanced technologies dedicated to steel scrap recycling.

It has been noted that the composition of steel scraps in the post-consumer category is of relatively lower purity [42, 43]. This adds to the challenge in the utilization of such scraps, as there are chances of no countermeasures during the sorting of these scraps. The scrap rate projected for 2050 is ~75% for the European Union. Similarly, for the United Kingdom, material flow analysis has been described by Davis et al. [44]. Most of the steel reutilization is observed in construction and automobile industries.

As noted in the previous section, the EAF steelmaking is the chief method of alloy steels. The alloying elements such as molybdenum, manganese, chromium, nickel etc. present in the scraps are of immense value, however during recycling of steel scraps the emphasis is on steelmaking only [45]. This leads to the loss of these important alloying elements to waste streams. Fair utilization of these alloying elements from these scraps can be highly contributing to the circularity of these elements.

With an overview of the key challenges in metal recycling, it is essential to examine some of the modern alternatives to the conventional metal reclamation routes. The final section of this chapter describes the modern advances in metal reclamation and recovery. Some of these processes are laboratory experiments, whereas some have been implemented to a higher scale.

5.6 MODERN METAL WASTE MANAGEMENT

As discussed in the previous sections, metal recovery from various sources depends on the raw feed of the process and the route of metal recovery. There are numerous challenges associated with the waste management and safe disposal. These have been described previously. In the final section, some newer methods are highlighted to study various aspects of waste management.

Van den Eynde et al. have discussed the possibility of using Laser Induced Breakdown Spectroscopy for the sorting of aluminum scraps [46]. It is reported that using this technology, the "twitch" fraction of aluminum shredded scrap can be properly sorted. The use of XRF has led to improved sorting of the scrap wastes. Similarly, Díaz-Romero et al. have included deep learning process into the LIBS sorting technology to enhance the efficiency of aluminum scrap sorting [47]. The intended application of these technologies is to reduce metal waste and improve the sorting of aluminum alloys, thereby increasing wrought alloy recycling.

Scrap sheet utilization for joint making has also been reported as one of the alternative methods of using scraps. As described by Ouyang et al., scrap aluminum sheets can be used as auxiliary materials for the joint making by punching the sheets to be joined along with the scrap metal sheet [48]. This leads to interlocking of the sheets and it enhances the deformation resistance of the joint.

To use the Non-Metallic Product (NMP), D. Bajare et al. have developed a method for the production of lightweight clay aggregate by milling NMP with clay followed by sintering [49]. Bonnin et al. have applied bi-objective optimizations using parameters like production rate vs energy demands, production vs operation cost and production vs environmental impacts [50].

Investigations to understand the changes in mechanical properties of the finished alloy/product after recycling of scraps or other metallic wastes are also necessary. Many researchers have focused on material characterization after the completion of the recycling process. It has been reported by Srinivasan et al. that the presence of intermetallic phases in the recycled AZC1231 magnesium alloy provided higher wear resistance, compared to the conventional alloy [51]. Shu-yan et al. have examined the mechanical properties of AZ31B magnesium alloy and reported that hot pressed recycled alloys had lower porosity indicating higher ductility [52]. Ductility is also improved when a double extrusion process is applied. Due to various compressive and shear forces applied during second extrusion, oxides are broken and dispersed into the matrix of the alloy, leading to higher ductility. Variation of chip size has also been investigated for similar solid-state recycling processes [53]; it is reported that the strength of the recycled alloy increases with total surface area of the chips. Similarly, low-temperature experiments have been conducted for AZ91D recycled alloys and it has been found that at -130°C, 360 MPa ultimate tensile strength is achieved [54].

Ghezzi et al. have explored the development of a predictive model for the recycling of steel waste [55]. Different categories of steel scraps and the impurities associated with these scraps have been incorporated into the new implicit-explicit dual model, making it a more comprehensive approach to the conventional non-dual formulated models. Similarly, steel scraps have been the subject of deep learning as the categorization of scraps has been a challenge for efficient metal recovery [56]. The determination of a scrap category by visual inspection has been training and modeled using the database collected and optimized. The aim of seamless scrap sorting can be achieved in the near future when the models of sorting are applied to larger scales.

With a carbon border adjustment mechanism (CBAM) policy, the European Union aims to reduce the generation of greenhouse gases. Correlating the CBAM policy to modern steelmaking, Rossetto has investigated the impacts of resource streaming on steelmaking [57]. Reduction of greenhouse gas emissions can be aimed at proper steel waste recycling.

The present chapter focused on the metal recovery from various metal-bearing sources. Utilization of scraps for metal recovery and understanding mining and beneficiation wastes as resources for metal and material recovery were important aspects described. Various metal sources for different metals have been discussed, along with some efficient methods for recovery. Important challenges, including environmental and economic constraints, have also been described. Some modern methods for metal recovery have also been presented to enhance the understanding of metal recovery.

The need of the present day is to focus on the important aspects of the environmental burden due to the adverse impacts of improper waste disposal. Metallurgical industries are one of the most crucial sectors that can advance towards greener production, leading to lower carbon footprint and CO_2 emissions. A circular economy coupled with environmental protection policies will enhance the possibilities of efficient metal recycling from these metal-rich wastes.

Appreciation of conventional metal production routes is also essential, as existing infrastructure for metal production is directly used for metal recovery from such wastes. Modernization and adaptation of these production routes is a possible way for improving metal recovery in industries.

The constraints of resources at the disposal of humans have led to the exploration of recycling routes. With the growing demand for better production and high-quality finished products, the industries are more oriented toward increasing metal production, whereas many environmental regulations have been put forth by government bodies across the world. This implies that constraints on the metal industries have increased manifolds. Adapting modern technologies for metal recovery from metal-rich waste resources thus becomes crucial.

REFERENCES

[1] Bian Z, Miao X, Lei S, et al. The challenges of reusing mining and mineral-processing wastes. *Science (80)* 2012; 337: 702–704.

[2] Whitworth AJ, Forbes E, Verster I, et al. Review on advances in mineral processing technologies suitable for critical metal recovery from mining and processing wastes. *Clean Eng Technol* 2022; 7: 100451.

[3] Shi J, Huang W, Han H, et al. Pollution control of wastewater from the coal chemical industry in China : Environmental management policy and technical standards. *Renew Sustain Energy Rev* 2021; 143: 110883.

[4] Hammond AA. Mining and quarrying wastes: A critical review. *Eng Geol* 1988; 25: 17–31.

[5] Ally AN, Blanche MM, Pettang Nana UJ, et al. Recovery of mining wastes in building materials : A review. *Open J Civ Eng* 2021; 11: 379–397.

[6] Capasso I, Lirer S, Flora A, et al. Reuse of mining waste as aggregates in fly ash-based geopolymers. *J Clean Prod* 2019; 220: 65–73.

[7] Abdel-Gawwad HA, Mohamed SA, Mohammed MS. Recycling of slag and lead-bearing sludge in the cleaner production of alkali activated cement with high performance and microbial resistivity. *J Clean Prod* 2019; 220: 568–580.

[8] Institute IA. *Material flow analysis A look at the numbers globally and in Asia.* 2023, The International Aluminium Institute. https://archive.international-aluminium.org/resource/material-flow-analysis-a-look-at-the-numbers-globally-and-in-asia/.

[9] Van den Eynde S, Bracquené E, Diaz-Romero D, et al. Forecasting global aluminium flows to demonstrate the need for improved sorting and recycling methods. *Waste Manag* 2022; 137: 231–240.

[10] Raabe D, Ponge D, Uggowitzer PJ, et al. Making sustainable aluminum by recycling scrap: The science of "dirty" alloys. *Prog Mater Sci* 2022; 128, 100947. DOI: 10.1016/j.pmatsci.2022.100947

[11] Sevigné-Itoiz E, Gasol CM, Rieradevall J, et al. Environmental consequences of recycling aluminum old scrap in a global market. *Resour Conserv Recycl* 2014; 89: 94–103.

[12] Hatayama H, Daigo I, Matsuno Y, et al. Evolution of aluminum recycling initiated by the introduction of next-generation vehicles and scrap sorting technology. *Resour Conserv Recycl* 2012; 66: 8–14.

[13] Zhu Y, Chappuis LB, De Kleine R, et al. The coming wave of aluminum sheet scrap from vehicle recycling in the United States. *Resour Conserv Recycl* 2021; 164: 105208.

[14] Stewart DJC, Barron AR. Pyrometallurgical removal of zinc from basic oxygen steelmaking dust – A review of best available technology. *Resour Conserv Recycl* 2020; 157: 104746.

[15] Wang X, Zhong Y, Kang Y, et al. Promoted acid leaching of Zn from hazardous zinc-containing metallurgical dusts : Focusing on transformation of Zn phases in selective reduction roasting. *Process Saf Environ Prot* 2022; 163: 353–361.

[16] Guresin N, Topkaya YA. Dechlorination of a zinc dross. *Hydrometallurgy* 1998; 49: 179–187.

[17] Ramadan A, Shash AY, El-Mahallawi IS, et al. Identification of copper precipitates in scrap based recycled low carbon rebar steel. *Mater Des* 2017; 120: 157–169.

[18] Wan B, Li W, Liu F, et al. Determination of fluoride component in the multifunctional refining flux used for recycling aluminum scrap. *J Mater Res Technol* 2020; 9: 3447–3459.

[19] Xiao Y, Reuter MA. Recycling of distributed aluminium turning scrap. *Miner Eng* 2002; 15: 963–970.

[20] Lin R, Liu B, Zhang J, et al. Microstructure evolution and properties of 7075 aluminum alloy recycled from scrap aircraft aluminum alloys. *J Mater Res Technol* 2022; 19: 354–367.
[21] Amini Mashhadi H, Moloodi A, Golestanipour M, et al. Recycling of aluminium alloy turning scrap via cold pressing and melting with salt flux. *J Mater Process Technol* 2009; 209: 3138–3142.
[22] Gómez F, Guzmán JI, Tilton JE. Copper recycling and scrap availability. *Resour Policy* 2007; 32: 183–190.
[23] Meskers CEM, Xiao Y, Boom R, et al. Evaluation of the recycling of coated magnesium using exergy analysis. *Miner Eng* 2007; 20: 913–925.
[24] Diener DL, Tillman AM. Scrapping steel components for recycling – Isn't that good enough? Seeking improvements in automotive component end-of-life. *Resour Conserv Recycl* 2016; 110: 48–60.
[25] Berlin D, Feldmann A, Nuur C. Supply network collaborations in a circular economy: A case study of Swedish steel recycling. *Resour Conserv Recycl* 2022; 179: 106112.
[26] Wang M, Chen W, Zhou Y, et al. Assessment of potential copper scrap in China and policy recommendation. *Resour Policy* 2017; 52: 235–244.
[27] Tian X, Zheng J, Hu L, et al. Impact of China's waste import policy on the scrap copper recovery pattern and environmental benefits. *Waste Manag* 2021; 135: 287–297.
[28] Dong D, Espinoza LAT, Loibl A, et al. Scenarios for anthropogenic copper demand and supply in China: Implications of a scrap import ban and a circular economy transition. *Resour Conserv Recycl* 2020; 161, 104943. DOI: 10.1016/j.resconrec.2020.104943
[29] Liu S, Zhang Y, Su Z, et al. Recycling the domestic copper scrap to address the China's copper sustainability. *J Mater Res Technol* 2020; 9: 2846–2855.
[30] Wang J, Ju Y, Wang M, et al. Scenario analysis of the recycled copper supply in China considering the recycling efficiency rate and waste import regulations. *Resour Conserv Recycl* 2019; 146: 580–589.
[31] Loibl A, Tercero Espinoza LA. Current challenges in copper recycling: Aligning insights from material flow analysis with technological research developments and industry issues in Europe and North America. *Resour Conserv Recycl* 2021; 169: 105462.
[32] Wang C, Huang X, Lim MK, et al. Mapping the structural evolution in the global scrap copper trade network. *J Clean Prod* 2020; 275: 122934.
[33] Hu X, Wang C, Lim MK, et al. Characteristics of the global copper raw materials and scrap trade systems and the policy impacts of China's import ban. *Ecol Econ* 2020; 172: 106626.
[34] Li J, Eheliyagoda D, Geng Y, et al. Examining the influence of copper recycling on prospective resource supply and carbon emission reduction. *Fundam Res* 2022. DOI: 10.1016/j.fmre.2022.09.022
[35] Ciacci L, Fishman T, Elshkaki A, et al. Exploring future copper demand, recycling and associated greenhouse gas emissions in the EU-28. *Glob Environ Chang* 2020; 63: 102093.
[36] Liu S, Liu W, Tan Q, et al. The impact of China's import ban on global copper scrap flow network and the domestic copper sustainability. *Resour Conserv Recycl* 2021; 169: 105525.
[37] Park H, Nam K, Lee J. Lessons from aluminum and magnesium scraps fires and explosions: Case studies of metal recycling industry. *J Loss Prev Process Ind* 2022; 80: 104872.

[38] Meshram A, Jain A, Rao MD, et al. From industrial waste to valuable products: Preparation of hydrogen gas and alumina from aluminium dross. *J Mater Cycles Waste Manag* 2019; 0: 0.
[39] Meshram A, Singh KK. Recovery of valuable products from hazardous aluminum dross: A review. *Resour Conserv Recycl* 2018; 130: 95–108.
[40] Golroudbary SR, Makarava I, Repo E, et al. Magnesium life cycle in automotive industry. *Procedia CIRP* 2022; 105: 589–594.
[41] Liu Y, Cui M, Gao X. Building up scrap steel bases for perfecting scrap steel industry chain in China: An evolutionary game perspective. *Energy* 2023; 278: 127742.
[42] Dworak S, Rechberger H, Fellner J. How will tramp elements affect future steel recycling in Europe? – A dynamic material flow model for steel in the EU-28 for the period 1910 to 2050. *Resour Conserv Recycl* 2022; 179: 106072.
[43] Dworak S, Fellner J. Steel scrap generation in the EU-28 since 1946 – Sources and composition. *Resour Conserv Recycl* 2021; 173: 105692.
[44] Davis J, Geyer R, Ley J, et al. Time-dependent material flow analysis of iron and steel in the UK. Part 2. Scrap generation and recycling. *Resour Conserv Recycl* 2007; 51: 118–140.
[45] Ohno H, Matsubae K, Nakajima K, et al. Toward the efficient recycling of alloying elements from end of life vehicle steel scrap. *Resour Conserv Recycl* 2015; 100: 11–20.
[46] Den Eynde S Van, Diaz-Romero DJ, Engelen B, et al. Assessing the efficiency of Laser-Induced Breakdown Spectroscopy (LIBS) based sorting of post-consumer aluminium scrap. *Procedia CIRP* 2022; 105: 278–283.
[47] Díaz-Romero D, Van den Eynde S, Zaplana I, et al. Classification of aluminum scrap by laser induced breakdown spectroscopy (LIBS) and RGB + D image fusion using deep learning approaches. *Resour Conserv Recycl* 2023; 190, 106865. DOI: 10.1016/j.resconrec.2023.106865
[48] Ouyang X, Zhang H, Chen C. Recycling of scrap sheet material: Enhanced joining mechanical properties of aluminum alloys by a novel auxiliary sheet flat-clinching process. *J Mater Res Technol* 2023; 25: 4609–4621.
[49] Bajare D, Korjakins A, Kazjonovs J, et al. Pore structure of lightweight clay aggregate incorporate with non-metallic products coming from aluminium scrap recycling industry. *J Eur Ceram Soc* 2012; 32: 141–148.
[50] Bonnin M, Azzaro-Pantel C, Domenech S, et al. Multicriteria optimization of copper scrap management strategy. *Resour Conserv Recycl* 2015; 99: 48–62.
[51] Bala Srinivasan P, Blawert C, Dietzel W. Dry sliding wear behaviour of a conventional and recycled high pressure die cast magnesium alloys. *Mater Charact* 2009; 60: 843–847.
[52] WU S yan, JI Z sheng, RONG S fan, et al. Microstructure and mechanical properties of AZ31B magnesium alloy prepared by solid-state recycling process from chips. *Trans Nonferrous Met Soc China (English Ed)* 2010; 20: 783–788.
[53] Wu S, Ji Z, Zhang T. Microstructure and mechanical properties of AZ31B magnesium alloy recycled by solid-state process from different size chips. *J Mater Process Technol* 2009; 209: 5319–5324.
[54] Li DH, Hu ML, Wang HB, et al. Low temperature mechanical property of AZ91D magnesium alloy fabricated by solid recycling process from recycled scraps. *Trans Nonferrous Met Soc China (English Ed)* 2011; 21: 1234–1240.
[55] Ghezzi A, Messerer F, Balocco J, et al. An implicit and explicit dual model predictive control formulation for a steel recycling process. *Eur J Control* 2023; 74, 100841.
[56] Xu W, Xiao P, Zhu L, et al. Classification and rating of steel scrap using deep learning. *Eng Appl Artif Intell* 2023; 123: 106241.

[57] Rossetto D. The carbon border adjustment mechanism: What does it mean for steel recycling? *Sustain Horizons* 2023; 5: 100048.
[58] Gottmyers Melwyn J, Chandragandhi B, Sathiyaseelan G, et al. Aluminium scrap recycling in a production furnace: Minimizing dross formation for sustainable and efficient recovery. *Mater Today Proc.* 2023. DOI: 10.1016/j.matpr.2023.05.340
[59] Baffari D, Buffa G, Ingarao G, et al. Aluminium sheet metal scrap recycling through friction consolidation. *Procedia Manuf* 2019; 29: 560–566.
[60] Sun B, Sun N, Wang L, et al. Efficient purification of scrap 1060 aluminum alloys contaminated with Fe and Si by super-gravity separation. *Process Saf Environ Prot* 2023; 177: 486–495.
[61] Sun B, Sun N, Wang L, et al. A new process for efficient purification of 6061 aluminum alloy scrap under semi-solid and super-gravity conditions. *J Clean Prod* 2023; 402: 136825.
[62] Samuel M. A new technique for recycling aluminium scrap. *Journal of Materials Processing Technology* 2003; 135: 117–124.
[63] Whalen S, Taysom BS, Overman N, et al. Porthole die extrusion of aluminum 6063 industrial scrap by shear assisted processing and extrusion. *Manuf Lett* 2023; 36: 52–56.
[64] Whalen S, Overman N, Taysom BS, et al. Effect of high iron content on direct recycling of unhomogenized aluminum 6063 scrap by shear assisted processing and extrusion. *J Manuf Process* 2023; 97: 115–124.
[65] Liu G, Wang Y, Fan Z. A physical approach to the direct recycling of Mg-alloy scrap by the rheo-diecasting process. *Mater Sci Eng A* 2008; 472: 251–257.
[66] Fechner D, Blawert C, Hort N, et al. Development of a magnesium secondary alloy system for mixed magnesium post-consumer scrap. *Mater Sci Eng A* 2013; 576: 222–230.
[67] Dudek P, Piwowońska J, Polczyk T. Microstructure of castings made of magnesium alloys based on recycling of the circulating scrap. *J Mater Res Technol* 2021; 14: 2357–2364.
[68] Uan JY, Lin MC, Cho CY, et al. Producing hydrogen in an aqueous NaCl solution by the hydrolysis of metallic couples of low-grade magnesium scrap and noble metal net. *Int J Hydrogen Energy* 2009; 34: 1677–1687.
[69] Yu SH, Uan JY, Hsu TL. Effects of concentrations of NaCl and organic acid on generation of hydrogen from magnesium metal scrap. *Int J Hydrogen Energy* 2012; 37: 3033–3040.
[70] Tzamtzis S, Zhang H, Xia M, et al. Recycling of high grade die casting AM series magnesium scrap with the melt conditioned high pressure die casting (MC-HPDC) process. *Mater Sci Eng A* 2011; 528: 2664–2669.
[71] Chang YC, Uan JY. Mg17Al12 phase in magnesium alloy waste facilitating the Ni2+ reduction in nickel plating wastewater. *J Hazard Mater* 2021; 403: 123556.
[72] Yam BJY, Le DK, Do NH, et al. Recycling of magnesium waste into magnesium hydroxide aerogels. *J Environ Chem Eng* 2020; 8: 104101.
[73] Lucci R, Padilla RL, Cantero S, et al. Refining of AZ91 magnesium alloy obtained in machining chips recycling. *Procedia Mater Sci* 2015; 8: 886–893.
[74] Hu M, Ji Z, Chen X, et al. Effect of chip size on mechanical property and microstructure of AZ91D magnesium alloy prepared by solid state recycling. *Mater Charact* 2008; 59: 385–389.
[75] Hu ML, Ji ZS, Chen XY. Effect of extrusion ratio on microstructure and mechanical properties of AZ91D magnesium alloy recycled from scraps by hot extrusion. *Trans Nonferrous Met Soc China (English Ed)* 2010; 20: 987–991.

[76] Hu ML, Ji ZS, Chen XY, et al. Solid-state recycling of AZ91D magnesium alloy chips. *Trans Nonferrous Met Soc China (English Ed)* 2012; 22: s68–s73.
[77] Li R, Wang L, Yang B, et al. Recovery of AZ91D magnesium alloy waste by vacuum gasification – Directed condensation. *Vacuum* 2023; 212: 112048.
[78] Scharf C, Ditze A. Recycling of black dross containing rare earths originating from melting and recycling of magnesium alloys. *Hydrometallurgy* 2015; 157: 140–148.
[79] Zhang L, Song Q, Liu Y, et al. An integrated capture of copper scrap and electrodeposition process to enrich and prepare pure palladium for recycling of spent catalyst from automobile. *Waste Manag* 2020; 108: 172–182.
[80] Liu W, Ma B, Zhou Z, et al. Efficient separation of impurities in scrap copper by sulfurization-vacuum distillation. *Vacuum* 2022; 202: 111145.
[81] Murali A, Plummer MJ, Shine AE, et al. Optimized bioengineered copper recovery from electronic wastes to increase recycling and reduce environmental impact. *J Hazard Mater Adv* 2022; 5: 100031.
[82] An J, Xie G, Xia W, et al. Preparation of fine copper powders by galvanostatic regime of electrolysis of copper scrap in a cylindrical electrochemical cell. *Powder Technol* 2021; 386: 193–198.
[83] Liu S, Cai Y, Zhang Y, et al. Selective separation of base metals and high-efficiency enrichment of precious metals from scrap copper anode slime. *Sep Purif Technol* 2022; 296: 121378.
[84] Wang J, Zhang Y, Cui K, et al. Pyrometallurgical recovery of zinc and valuable metals from electric arc furnace dust–A review. *J Clean Prod* 2021; 298: 126788.
[85] Tang H, Peng Z, Wang L, et al. Facile synthesis of zinc ferrite as adsorbent from high - zinc electric arc furnace dust. *Powder Technol* 2022; 405: 117479.
[86] Liang M, Di H, Hong Y, et al. Inhibition of Fe 4 Ge 3 O 12 formation in the leaching process of zinc oxide dust containing germanium by ultrasonic and iron powder. *Arab J Chem* 2023; 16: 104794.
[87] Rudnik E. Hydrometallurgical recovery of zinc from industrial hot dipping top ash. *Trans Nonferrous Met Soc China* 2020; 30: 2239–2255.
[88] Rudnik E. Recovery of zinc from zinc ash by leaching in sulphuric acid and electrowinning. *Hydrometallurgy* 2019; 188: 256–263.
[89] Anandaraj S, Karthik S, Vijaymohan S, et al. Effects of using white flour, zinc oxide and zinc ash as an admixture in mortar and concrete. *Mater Today Proc* 2022; 52: 1788–1793.
[90] Dikici T, Demirci S. In fl uence of thermal oxidation temperature on the microstructure and photoelectrochemical properties of ZnO nanostructures fabricated on the zinc scraps. *J Alloys Compd* 2019; 779: 752–761.
[91] Sinha S, Choudhari R, Shekhar S, et al. Materials today: Proceedings recovery of zinc as zinc phosphate and by-products from hydrometallurgical treatment of hot-dip galvanizing (Zn-Al) dross. *Mater Today Proc* 2021; 46: 1532–1536.
[92] Sinha S, Choudhari R, Mishra D, et al. Valorisation of waste galvanizing dross : Emphasis on recovery of zinc with zero effluent strategy. *J Environ Manage* 2020; 256: 109985.
[93] Wang Z, Gao J, Shi A, et al. Recovery of zinc from galvanizing dross by a method of super-gravity separation. *J Alloys Compd* 2018; 735: 1997–2006.
[94] Prasad S. Zinc powder preparation from zinc dross. *Mater Today Proc* 2022; 67: 336–341.
[95] Huajun Z, Zhenghai G, Yunpeng Z. Electrorefining zinc dross in ammoniacal ammonium chloride system. *Hydrometallurgy* 2008; 90: 8–12.

[96] Barakat MA. Recovery of metal values from zinc solder dross. *Waste Manag* 1999; 19: 503–507.
[97] Chen C, Ma J, Li Y, et al. Recovery of tin from tin-coated copper-clad steel wire scrap using surface sulfuration-vacuum volatilization. *Vacuum* 2023; 212: 112052.
[98] Shewalul YW. Experimental study of the effect of waste steel scrap as reinforcing material on the mechanical properties of concrete. *Case Stud Constr Mater* 2021; 14: e00490.
[99] Liem-Nguyen V, Sjöberg V, Dinh NP, et al. Removal mechanism of arsenic (V) by stainless steel slags obtained from scrap metal recycling. *J Environ Chem Eng* 2020; 8: 103833.
[100] Huy DH, Seelen E, Liem-Nguyen V. Removal mechanisms of cadmium and lead ions in contaminated water by stainless steel slag obtained from scrap metal recycling. *J Water Process Eng* 2020; 36: 101369.

6 Electronic Waste

Arunabh Meshram and Krishanu Biswas

6.1 INTRODUCTION

In the previous chapters, it has been shown that metallurgical wastes have great potential to become reliable sources of metals, ceramic. The methods of metal recovery have advanced extensively, and this advance has encouraged many researchers to address the issue of electronic waste recycling. While metallurgical wastes are generated in the industrial sectors, electronic wastes are generated at various stages of applications. With the growing demand for newer and advanced electronic devices and appliances, the production of electronic goods has escalated manifolds. This has, in turn, led to a large-scale pileup of discarded electronic devices.

The ever-advancing electronics industry sector produces a wide variety of products for the consumers. The old and obsolete electronic devices are discarded and disposed, even if the devices are operational and have not reached their end-of-life. Steep advances in technologies have led to quicker obsoletion of functional devices and appliances. Statistical insights into electronic waste have been presented by many researchers and the growing concern regarding its environmental impacts has been alarming. It is estimated that 57.4 million tons of electronic waste have been generated in the year 2021[1]. This trend of waste generation is expected to rise further [2].

Emphasis on metal recovery from electronic waste has become especially evident, now that the natural resources for the continued production of electronic devices are limited. Many crucial elements necessary for key applications in electronic devices are very rare. Due to limited resources, the cost of production increases. Understanding the loop of material cycle is becoming even more important for continued production of high-quality electronic devices. If the raw materials cannot be made available due to shortage of conventional ores, the demand needs to be balanced with the secondary sources of metals/materials. This clearly indicates that industrialists need to shift their focus from conventional material production routes to recycling routes with E-waste stream being the primary source of metals for these specific applications. It is observed that many of these elements are applied for semiconductor or electronic applications and after recycling, these will be reused in similar fields.

DOI: 10.1201/9781003619680-6

Taking example of end-of-life photovoltaic modules, it has been estimated that if the recycling of the existing solar panel waste is not conducted properly, the future contribution of the PVMs will reach nearly 10% of the total electronic waste [3, 4]. This indicates that even solar panel wastes are also becoming significant contributors to the overall electronic waste burden.

This chapter focuses on the categorization of electronic wastes, understanding the economic benefits of recycling different waste streams, key challenges in the conventional material recycling processes and modern alternative methods of achieving higher metal recovery. While this chapter aims at presenting a holistic view of the electronic wastes, it is important to note that each stream of waste is equally important for achieving higher material recovery and ensuring reduction of environmental burden.

6.2 TYPES AND SOURCES OF ELECTRONIC WASTE

Appreciation of E-waste management is possible when a well-defined system of waste stream sorting is in place. Such sorting mechanisms will require state-of-the-art visual aid, coupled with mechanized sorting technology that is programmed for efficient detection and segregation of different categories of E-waste [5]. There are essentially six categories of electronic waste: large equipment, smaller household equipment, lighting lamps and other devices, screens and related devices, thermal and temperature maintaining (heating and refrigerating) devices and computing and communication devices. The categorization has been illustrated in Figure 6.1.

As noted in Figure 6.1, it is understood that the six categories of E-waste are merging into each other and many of the electronic devices fall into multiple categories. Taking examples within each category, it is seen that screens and display wastes have internal electronic circuits that can be categorized into small equipment or communication devices. Similarly, parts of discarded computers and laptops can have screens. This shows the complex nature of electronic wastes. Figure 6.1 shows the corresponding overall chemistry of E-waste. It is evident, they contain metals, alloys, ceramic as well as polymer. Hence, E-waste is usually considered as "urban mine". We can extract, recover, and recycle many of these metals, ceramics, and polymers by use of innovative technologies. Some of these technologies are conventional and some are modern and innovative. We shall discuss some of these technological innovations in the following.

A strict classification of the electronic waste stream is, therefore, not feasible. The key aim of the recycling route should be to efficiently recover valuables from these waste streams. Appreciation of materials present in the E-waste streams is very crucial for this. The next sections describe some examples of material recycling. Urban mining is a novel concept that accepts various waste streams as raw materials for valuable product development and material recovery. Unlike conventional mining and metallurgical operations, urban mining focuses on the utilization of resources already available after basic material processing. Valorization of materials in the wastes is necessary before developing a recycling route.

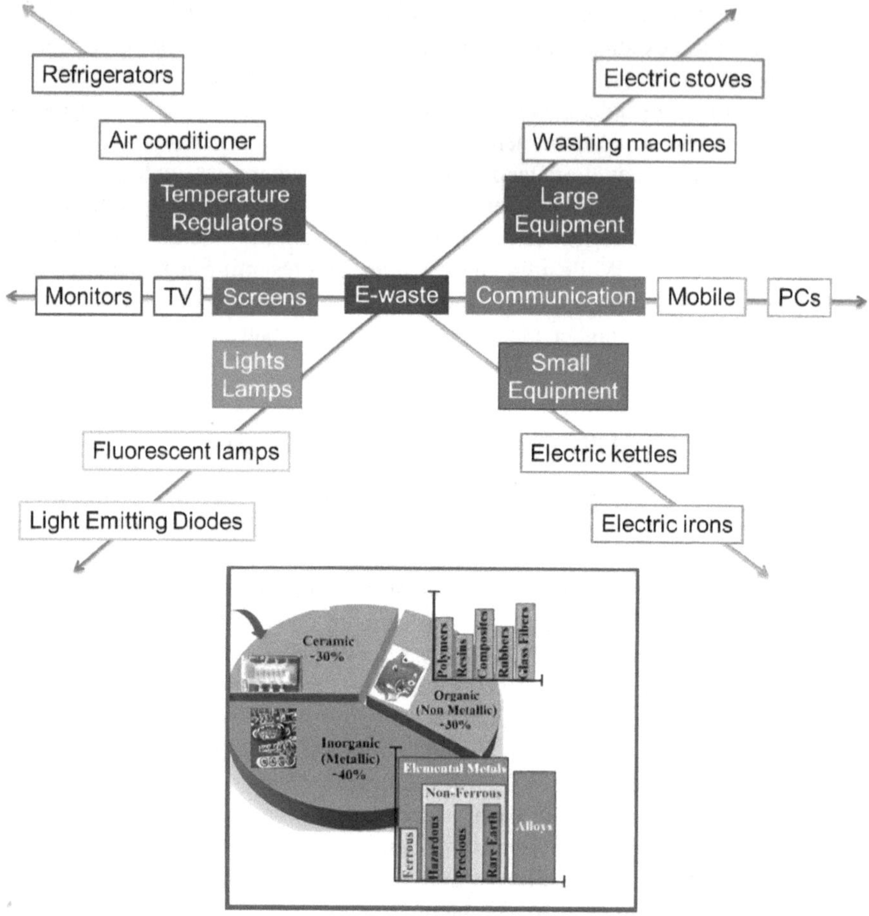

FIGURE 6.1 Categories of E-waste and some important electronic devices within these categories; typical composition of E-waste, showing different types of material present in it.

6.3 ELECTRONIC WASTE AS URBAN MINES

Similar to the metallurgical wastes described in the previous chapters, E-waste has now been highlighted as an important source of metallic waste. As noted in Table 6.1, the important elements present in different segments of E-waste, it is evident that with the variety of waste present in the E-waste stream, the concentration of these elements varies.

Electronic wastes have been brought into the material cycle, as the elemental composition of these wastes have become valuable in the recent years. As noted in the other sections, the cost of production of various electronic devices heavily depends on the availability of these elements. Efficient extraction of these elements from these wastes is therefore important, as these wastes are now considered as secondary resources. Many rare-earth elements like Tb, Eu and Y are present in fluorescent

TABLE 6.1
Variation of concentration of elements in different E-waste streams

Waste Stream	Cu	Fe	Al	Au	Ag	Zn	Co	Ni	Li	Mn	REE	Others
Waste Printed Circuit Boards	9.7–47.5	3–10.36	2.7–4.13	0.014–0.035	0.08–0.13	1–1.2	-	0.69–1.25	-	-	-	2–3.3 (Sn)
Li-ion batteries	10–17	-	6–7	-	-	-	5–20	5–10	5–7	6–19	-	-
Photovoltaic modules	0.11	-	0.53	0.05	-	-	-	-	-	-	-	0.05 (Pb and Sn)
Fluorescent lamps	-	-	-	-	-	-	-	-	-	-	0.4–1.3 (Tb) 1.1–1.9 (Eu) and 16.4–28 (Y)	-

Sources: [3, 6, 9–13].

lamps. These lamps are used in household and commercial scale and the discarded lamps need to be recycled for efficient metal recovery. Recovery of these rare-earth elements from the discarded lamps itself is a challenge. Dismantling of the lamps for the recovery of metals from phosphors is an important pretreatment step.

Noting the wide variety of elemental composition of these wastes, it is important to develop common sorting procedures to reduce the burden on the downstream operations. A categorization of wastes usually on the basis of type of electronic waste will ensure better pretreatment and physical-chemical-mechanical operations for higher metal/material recovery.

Valorizing these wastes and placing economic constraints on an expensive recycling process is another method of developing alternative methods of material processing. If recycling processes developed for material reusing/refurbishing/recovery are highly expensive or environmentally degrading, development of cost-effective and environment-friendly processes becomes necessary.

Taking example of the discarded photovoltaic modules (solar panels), it is understood that most of the weight of the solar panels is due to the glass sheet and metal frame. The electronic device encapsulated in the solar panel has all other important elements present in it. These elements make the solar panel functional and efficient for the conversion of solar energy to electricity. Silver is one such element. The economic contribution of silver to the production and functioning of solar panels is significant. Many researchers are focusing on the development of alternative recycling routes for efficient recovery of silver and metallurgical grade silicon for reusing in the photovoltaic modules.

While dismantling of the end-of-life photovoltaic modules is a big challenge, the internal electronic circuits get damaged and the recovery of silicon wafer is not achieved. For recovery of silicon wafers, it is important to remove the encapsulant polymer layers that are present in the solar panels. Recovery of other metals like copper, lead and silver can be done with different metallurgical processes combining hydrometallurgy, electrometallurgy etc.

Urban mining is achieved when materials from important waste streams are efficiently recovered. There are conventional concepts of ore comminution that can be brought into the material cycle. Some of these methods involving comminution are described in the next section. With the application of these beneficiation operations, the metal recovery is enhanced and the concentration of metals in the feed is increased. The discards of the physical operations include materials with lower economic importance or recovery/reusing of such materials is extremely difficult.

6.4 CONVENTIONAL ELECTRONIC WASTE BENEFICIATION TECHNIQUES

Most of the E-waste generated by human society is heavily polluted by other wastes generated. This hampers the recycling process. The collective stream of the waste is usually not available for material recycling and recovery. In order to increase metal recovery from the available waste streams, it is important to improve the norms for material sorting before they are brought for material recovery.

The most important step for material recovery, therefore, is material sorting and pretreatment/ beneficiation to increase the metallic content in the feed. As most of the electronic waste streams have other materials associated with them, it is advantageous to have proper pretreatment to enhance the metal recovery. For instance, recycling waste printed circuit boards of various electronic devices like discarded motherboards of PCs, laptops, cellphones, washing machines etc. involves a large variety of materials, including waste plastics, polymers, composites, metals, and some ceramics. Soldering of conducting components to the circuit boards is also present in the waste printed circuit boards. To efficiently commence the metal extraction from various materials, present in these streams, proper pretreatment is important.

Given the versatility of materials in the waste streams, a combination of physical, mechanical, and chemical operations may be necessary for the initial pretreatment and beneficiation. Comparison of electronic wastes with metallurgical wastes provides an important insight into the necessity of pretreatments. Most of the metallurgical waste streams may have heterogeneous composition within a stream, but the streams usually have a single type of waste. E-waste, on the other hand, does not ensure a fixed type of waste in the stream. Even for a single type of waste, the composition of the waste can vary depending on the company of origin, service environment and year of manufacture. With such a wide variety of parameters, it is very challenging to ensure a stream of fixed composition for metal recovery.

As mentioned previously, the operations essential for material separation are broadly classified as physical, chemical, and mechanical operations. Mechanical operations involve the use of various conventional comminution processes and facilities like crushers, grinders, and physical separators, like sieve shakers. These mechanical operations are responsible for the physical separation of materials after primary comminution. There are some methods like heavy media separation and density separation that enhance the material recovery in the comminution product. Fine particles of waste can be subjected to various media separation. Physical operations involve sorting based on visuals, like manual sorting, manual separation of various components of the waste. While mechanical and physical operations can be conducted simultaneously, chemical operations need separate investment, as these involve the use of chemicals like acids, alkalis etc. for the separation of impurity materials/metals.

Chemical operations can involve water washing, acid washing or alkali washing, depending on the composition of the feed and the desired composition of the metal extracted from the waste. Some important metals that are recovered at the beginning of metal extraction from electronic wastes are base metals like copper, aluminum, and iron. These metals are present in many E-waste streams and are easily recovered by using chemical processes. After the recovery of important metals from the waste, the secondary metals like gold, silver, platinum, palladium, nickel, manganese etc. can be economically recovered. For this, there are combinations of metallurgical processes (pyrometallurgical, hydrometallurgical and electrometallurgical) are employed for higher material recovery.

As observed in Figure 6.2, there are many pretreatment operations within the three categories mentioned. Each of the pretreatment provides a different set of finished

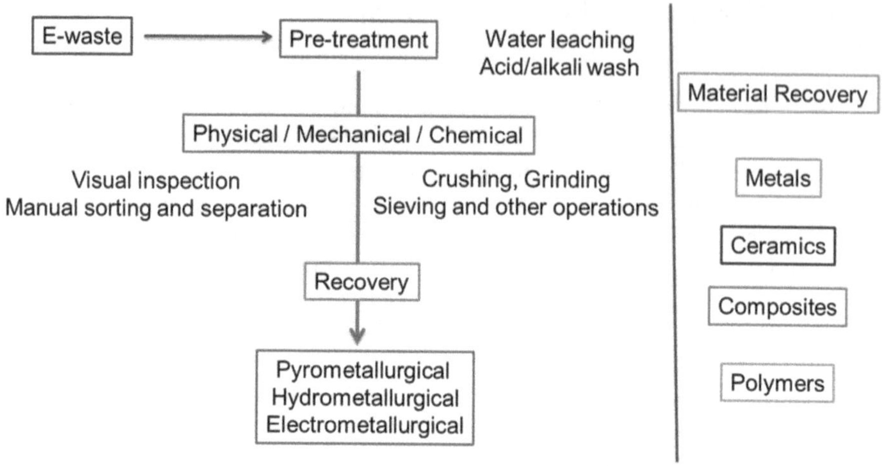

FIGURE 6.2 Application of various operations for material recovery from electronic waste.

products, depending on the pretreatment process. Finished products also bear certain characteristics on the basis of which they are classified or sorted, for example, some particles respond to external magnetism, whereas some particles do not. These operations are seldom carried out individually as many of these operations are interdependent.

Reduction of particle size for the recovery of valuable materials in the finer feed and increasing the metallic content in the finer feed are key steps that are usually followed by other physical or mechanical operations like heavy media separation etc. Combinations of the pretreatment operations simply increase the chances of metallic concentration and separation of non-metallic fraction for the waste streams.

Not just conventional pretreatment processes, but newer technologies involve delamination, separation based on physical properties like density, magnetism etc. are also included in the pretreatment segment. These technologies are very beneficial for enhancing metal fraction of the feed. Taking example of various delamination reagents like Dimethyl formamide (DMF) and Dimethyl acetamide (DMA), it has been reported that metal clads are easily recovered from the waste printed circuit board sandwiches [6, 7]. The epoxy resin between the layers reacts with the delamination reagents, leading to separation of glass fibers and metal clads. These metal clads can be further processed to extract metals like Cu, Au, Ag etc. [8]. Some other technologies involve ultrasonication of the feed to enhance the leaching efficiency.

Water washing, acidic/ alkaline washing to remove the dissolvable impurities in the initial stages is important. The pretreatment stages are used for recovering some dissolvable salts or metallic values that react with mild acidic and alkaline solutions. It should be noted that the actual hydrometallurgical processes like leaching for dedicated metal recovery will follow the feed that is washed and dissolvable impurities are removed.

The hydrometallurgical process mentioned after pretreatment is highly benefited by washing of E-waste streams. This is because many of the non-metallic fractions are removed for the feed after pretreatment. The consumption of acid/alkali for the recovery of metals is reduced significantly. The metal recovery improves as the leaching reagents are available for reaction and the partition of reagent is not observed.

Combination of pyrometallurgical, hydrometallurgical and electrometallurgical process is essential for the recovery of all categories of materials (Figure 6.2). Depending on the economic value, the recovery route is chosen between unit processes of these routes. Metals, ceramics, polymers, composites, and other classes of materials are extracted when pretreatment beneficiation is combined with recovery processes.

In the next section, some of the challenges related to metallurgical processes involved in recycling of E-waste streams are highlighted. It is important to note that the feed of the E-waste cannot be compared to conventional metallurgical wastes, as the complexity of the feed increases as a large number of materials are involved in producing electronic devices. After pretreatment, the chief methods are involved in material recovery. These are described in the next section.

6.5 ISSUES RELATED TO THE CONVENTIONAL TECHNIQUES

In the previous sections, the problem of E-waste has been thoroughly described. Important aspects of material beneficiation have been discussed, highlighting the advantages of comminution and particle size reduction and material sorting. These methods of material pretreatment have proven to be crucial for efficient metal recovery, but higher metal recoveries have been reported by following optimized methods.

This indicates that the pretreatment methods need to couple with proper routes for better metal recovery and reducing environmental burden. One of the most important hurdles in this path is faced in most of the developing countries in Asia, Africa, and other continents. The improper and unorganized collection of E-waste and its management is the biggest challenge faced by many governing bodies. This unorganized E-waste recycling method needs to be replaced by well-informed personnel, who will ensure proper metal recycling, maintaining negligible materials loss to environment, simultaneously reducing the burden on the environment. Figure 6.3 describes the key aspects of conventional metallurgical processes involved in completing the material cycle of E-waste streams.

Observing Figure 6.3, there are pyrometallurgical, hydrometallurgical and electrometallurgical processes involved in metal recovery from E-waste streams. As noted in the previous sections, the pretreatment leads to concentration of metallic feed, readily available for further processing. Most of the pyrometallurgical processes involve the smelting process, where the raw material feed is charged into a smelter unit and subjected to smelting operation in the presence of a reducing agent. This leads to partitioning of different elements into the metal phase and slag phase. Some elements are included in the metal phase, whereas the rest is present in the slag, depending on the characteristics of slag produced during the smelting operation.

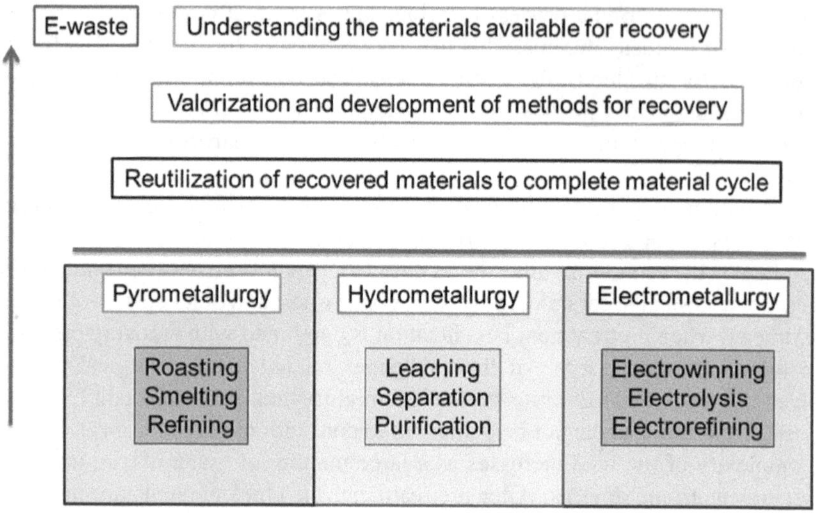

FIGURE 6.3 Combination of metallurgical processes for E-waste recycling.

Some volatile materials are also lost to the gaseous phase, like lithium from cathode materials of spent Li-ion batteries. The recovery of lithium is a challenge, as purification of flue gas of the smelter process requires additional investment. Volatile matter recovery poses a great challenge to pyrometallurgical processes, just like the separation of metallic values from the master alloy produced after smelting. In order to use the metallic content, further hydrometallurgical or electrometallurgical process is required for efficient separation.

When smelter process is followed directly after comminution of E-waste streams, the non-metallic fraction is also taken into the furnace. The smelting process is affected as the slag volume for such feed is higher and the metal recovery is affected by the presence of non-metallic fraction.

This is the chief motivation for understanding the materials involved in a given waste stream (Figure 6.3). A stream of spent batteries will have different set of materials, compared to waste printed circuit boards of mobile phones, computers and laptops. This indicates that every stream of waste should be thoroughly observed for developing the mechanisms of sorting and processing. If all different categories of wastes are directly processed (like smelted or leached), the effective metal recycling will be highly affected. Loss of metals to non-metallic fraction will be enormous. After proper identification of the materials, valorization is important. This step is the driving force for the development of dedicated recycling processes.

Recycling of single use plastic/polymer material may not be economically beneficial compared to silver or gold recovery for industries, although recycling plastics reduces environmental burden significantly. A trade-off between materials with economic impact and environmental impact is mostly seen for various waste streams. Therefore, a judicious decision should be made to arrive at a proper recycling route.

Combination of metallurgical processes like roasting, leaching and electrowinning/ electrolysis can be very beneficial for metal recovery. Most of the metals (apart from the base metals like copper, iron etc.) are present in significantly less concentration. With lower concentration of metallic values, hydrometallurgical processes like leaching, separation and purification are used for recovery. Solvent extraction and ion exchange are important processes used for the recovery of metallic values from leach liquor. Newer processes like polymer inclusion membranes are also being explored for increasing metal recovery from leach liquor containing multiple metal ions,

Taking the example of spent lithium-ion batteries, it is understood that depending on the cathode, there are various classes of Li-ion batteries. LCO type batteries usually have lithium and cobalt in cathode. The recovery of Li, Co and other elements from the cathode material is a challenge, as all the elements behave differently when exposed to leaching with various reagents like H_2SO_4, HCl or NaOH and KOH [9]. Some batteries also have protective sheets of steel or aluminum. These protective sheets are also exposed to leaching operations and their recovery is dependent on many parameters like particle size, concentration of reagent, agitation etc.

Using conventional processes like roasting and smelting, followed by fire-refining or electrorefining, many metals can be recovered successfully from E-waste streams. Reutilization of recovered metals/ metallic values is of crucial importance. Determination of the quality of metal/ products of recycling leads to its potential applications in various fields.

6.6 MULTIFACETED PROCESSING OF E-WASTE

As mentioned earlier, E-wastes have the lion's share in world of waste today. This is primarily because of the fact utilization of electronic devices are all pervasive across different countries, different strata of population and rich, middle class as well as poor. Hence, the concept of "Waste-to-Wealth" is always of tremendous interest for various reasons in the present context, more so for electronic waste, which has emerged as the fastest-growing waste stream in the world. In 2019, the world generated a record 53.6 million metric tons (Mt) of E-waste, a massive 21% increase in just five years [14]. The global generation of E-waste is estimated to reach 74 million Mt by 2030, almost doubling in tonnage in just 16 years and about 150 million Mt by 2100. Therefore, the effective utilization of large amounts of E-waste by recovering valuable materials *via* sustainable means and thus protecting the environment has been an urgent and imperative task. Though E-waste consists of waste electronic and electrical equipment of varying sizes, those with a small life span add the major weight to the growing pile of E-waste. This expendable equipment includes mobile phones, tabs, laptops, computers, hard disk, and other electronic accessories.

With a rapid boom in technological advancements and generic perception of newer is better, each new electronic device makes the older devices obsolete. As a result, there is a drastic reduction in the average lifespan of electronic goods, and a massive increment in end-of-life (EoL) electronic products, commonly known as E-waste. A recent survey shows that ~20–50 million tons of EE-Waste is discarded annually around the globe due to malfunctioning of equipment and outdated technologies. At present, statistical data [15–16]. outline different volumes of E-waste generation in varied corners of world and project its gigantic increase in near future, which is alarming for both environment and human. To subdue the problems imposed by E-waste increment, counter efforts are being made by various international organizations expeditiously. The enhanced digitalization of devices has equipped them with integral electronic circuit boards, generally termed as printed circuit boards (PCBs). These PCBs are a crucial part of all electrical and electronic products, accounting for ~3 wt.% of overall electronic component and are the most economically attractive portion of E-waste. Generally, three main categories of materials can be retrieved from PCBs[4], i.e. metals (~40%); polymers (~30%) and remaining ceramics/glasses[5].Further, metals in PCBs can be broadly categorized as base metals like Fe, Cu, Sn, and Al; rare metals like Er, Ce, Nd; noble metals like Au, Ag, Pd and hazardous metals like Cr, Pb, Cd. Usually, three major techniques are followed for E-waste recovery, i.e., physical, chemical, and biological processes [17–19]. Physical methods involve defragmentation of PCBs into small pieces followed by pyrolysis at elevated temperature for the recovery of precious metals such as Au and Ag [12]. In chemical processes, these metals are recovered by their selective etching, whereas, in biological processes, the precious metals are leached out from the electronic waste through living organisms [17–18]. After the extraction of desired metals, remaining E-waste is suitably disposed irrespective of the technique followed for the same. A recently developed technology by group led by the present authors [19]. demonstrating zero waste disposal (i.e.,100% recovery of component) through cryomilling has attracted significant scientific attention. The process involved separate recovery of metals, polymers, and oxides for their utilization in different applications. Further, recovery of metals in pure state and alloys for the extraction of individual constituent by additional extraction processes was also demonstrated. In this backdrop, it is important to establish easily scalable and green routes for beneficiation and practical usage of constituent materials from the components of E-waste. The waste printed circuit boards (PCBs) have 40% (by weight) metallic component, 30% polymeric component, and the remaining is ceramic residue. They can be treated easily and effectively to recover and recycle important metallic, ceramic as well as polymeric contents. One can even use the recovered materials for downstream production of other essential products and thereby increase the efficiency of use. However, one needs to use green and environmentally friendly routes so that no further waste is generated. We shall discuss the use of such routes for beneficiation of components from PCBs [19].

The waste PCBs can mechanically be segregated based on the size and density of the components. They are then broken down into small pieces of 4–6 mm size using a grinder. The dismantled PCBs are ball milled at -160±10°C, maintained using liquid nitrogen. Further, the components of PCBs are separated from each other by density separation of the milled powder, using repeated filtration and magnetic stirring in distilled water. While the heavier metallic residue settles at the bottom of the flask, the polymeric component remains at the top, which is subsequently removed and dried separately. This procedure can be repeated several times in order to obtain the maximum recovery of both metallic and polymeric powders. Figure 6.4 shows the detailed description of the process used in this route. The proposed process has shown the ability for near-net recovery of these constituents without using any non-biodegradable substances or chemicals.

Of the various pretreatment methods for the recovery of metallic components from PCBs, mechanical separation is one of the most promising, as it does not produce any chemical change in the content during processing. The basic steps involved in mechanical separation are screening of components as per their shape,

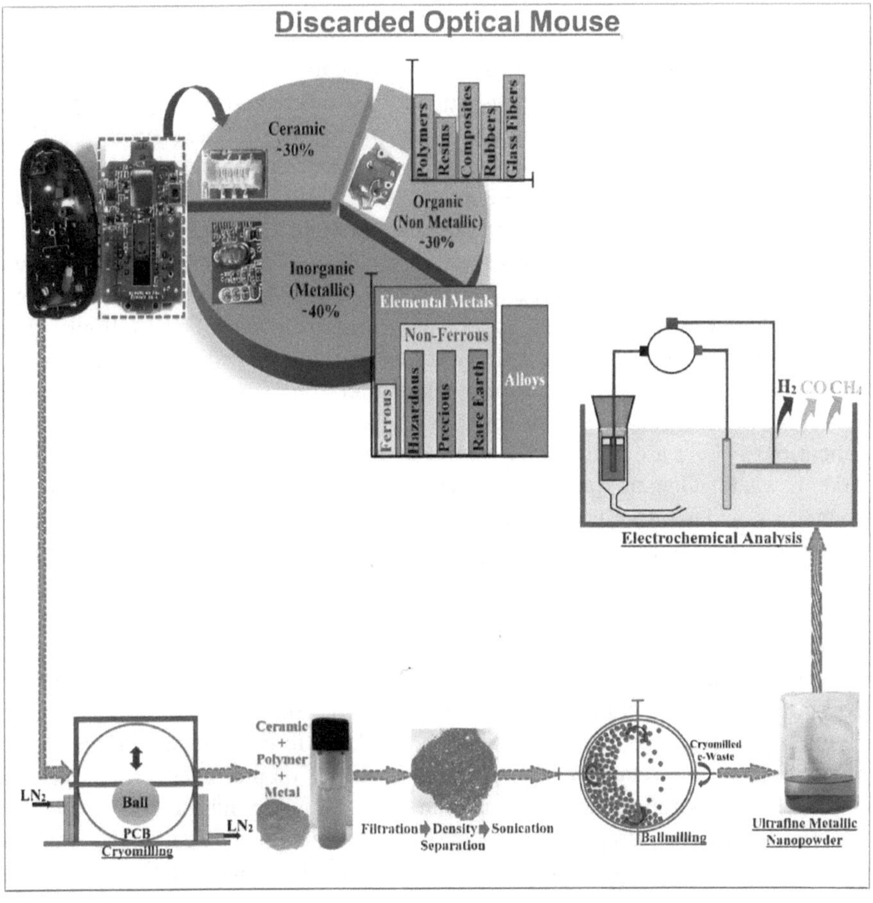

FIGURE 6.4 Green route to recycle and use metallic content of E-waste.

size, and density. Post crushing in mortar and pestle, the pieces of PCBs were initially milled using a custom-built Cryomill at ~-160±10°C; cooled using liquid nitrogen (LN_2) under Ar atmosphere for 7 h. The cryomilled powder was stored in methanol to avoid its oxidation. Afterwards, the metallic components present in PCBs were extracted by repeated filtration in distilled water *via* density separation from its polymeric and ceramic counterparts by repeated magnetic stirring, which was further followed by prolonged ultrasonication in acetone for the maximum dissolution of polymeric content. The metallic content can further be utilized for various purposes. Since the cryomilled and recovered materials contain nanostructured metals (Cu, Ni, Fe, Ag etc), we can use it as catalysts for hydrogen generation from various liquids (alcohol, toluene etc) and effectively make use of it. There are other uses of these metallic nanoparticles- mix and prepare alloys using melting-solidification or powder metallurgy route. In a nutshell, the recovered metallic portions can be used for various purposes.

6.7 RECYCLING POLYMERIC COMPONENT OF E-WASTE

The polymeric part can easily be recovered from the cryomilled powder by using physical separation/gravity separation. This can be used for synthesis of advanced materials; graphene or CNTs by using liquid phase laser ablation. Figure 6.5 shows the details of the such processing route and quality of graphene produced [20]. Across the globe, graphene is considered as wonder materials and it can be used for various purposes. This green, easily scalable, and sustainable method of synthesizing graphene from one of the major components of E-waste, i.e., the polymeric component, using pulsed laser ablation (PLA). The synthesis route shows great promise in synthesizing high-quality graphene. Microscopic investigations highlight that graphene has four layers with a high degree of crystallinity. The high-quality synthesized graphene has further been utilized for highly conductive electronics contacts, useful for future robust electronics applications. Similarly, CNTs produced from polymeric part of the E-waste can be used for various applications. There are other processing avenues for use of the polymeric portion of the E-waste. One of the notable uses involved compression molding of the polymeric powder into electrical switch boards. The extraction of the polymeric component from discarded printed circuit boards (PCBs) by employing a synergistic approach involving mechanical segregation and cryo-temperature milling and thereafter using compression molding to cast electrical wall switch casing from this polymer is the basis for such use. The PCBs used in typical electronic equipment contain a wide array of high-performance polymers, including FR-4, also known as flame retardant 4 (FR-4), polyimide, bismaleimide triazine (BT), polytetrafluoroethylene (PTFE), and polyethylene terephthalate (PET). These polymers provide exceptional properties in union, including electrical insulation, mechanical strength, and dimensional stability, while maintaining controlled dielectric properties for signal integrity. These polymers resist heat, and chemicals and offer adhesion for reliable connections. Keeping these in mind, these polymers can be used as potential candidates for electrical wall switch casings, which require

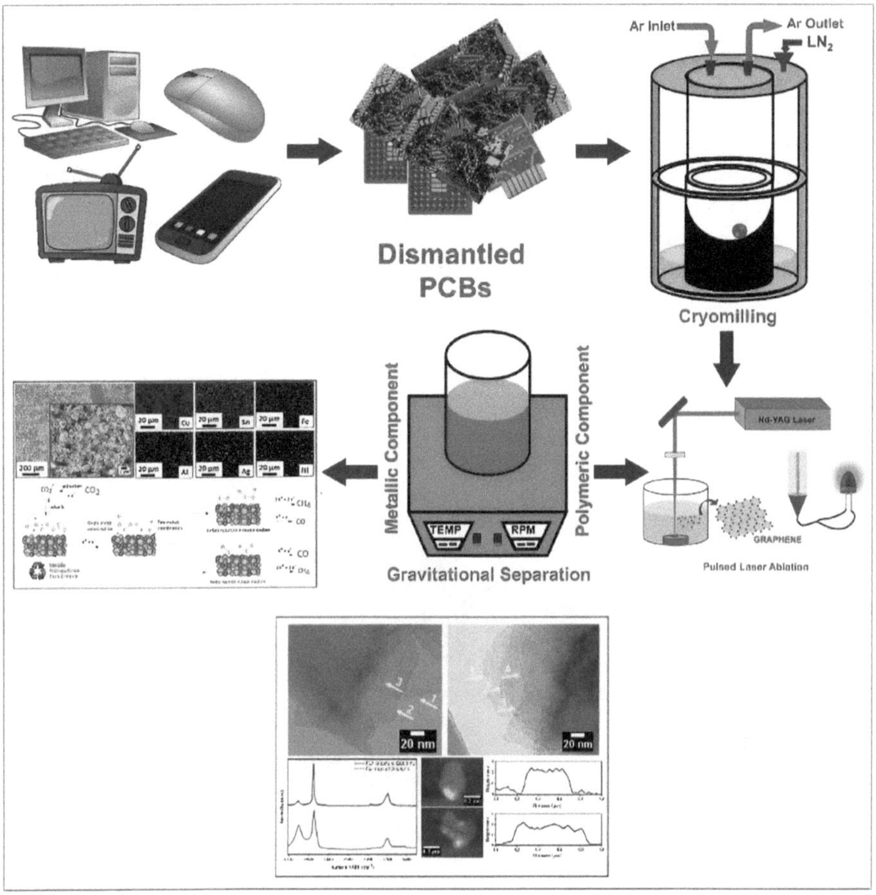

FIGURE 6.5 Green route to recycle and use polymeric content of E-waste for downstream green technologies.

electrical insulation, impact resistance, durability for repeated mechanical operation, and flame resistance for safety.

The first notable feature of the idea is the application of mechanical separation techniques in order to extract polymeric constituents from waste PCBs. The process entails mechanical disassembling the discarded PCBs in order to extract the metallic and ceramic components of greater weight while preserving the integrity of the polymeric component. The utilization of the mechanical separation technique not only exhibits environmental sustainability but also facilitates the retrieval of constituents without necessitating chemical interventions. The further novelty of this approach is in integrating mechanical separation techniques with cryo-temperature milling. Cryo-temperature milling entails exposing the components of waste PCBs to mechanical milling at extremely low temperatures, typically about -160±10°C [21]. This procedure is aided by creating an environment with an Argon (Ar) atmosphere and utilizing liquid nitrogen cooling. The milling process conducted at extremely low temperatures proves to be highly successful in the fragmentation of polymeric materials, resulting in their conversion to a powdered state. This transformation enhances the ease of handling and facilitates further processing operations. After cryo-temperature milling, the polymeric E-waste in powdered form undergoes density separation. This procedure entails the iterative process of filtration and magnetic stirring in distilled water. The underlying premise of this phase involves leveraging the disparities in density between the polymeric constituents and the denser metallic and ceramic constituents. Consequently, the polymeric constituents may be effectively segregated and accumulated, whilst the denser constituents precipitate and accumulate in the lowermost region of the collection vessel. Figure 6.6 schematically summarizes the proposed polymer extraction route. The utilization of technology in the manufacturing sector has become increasingly prevalent in recent years. The

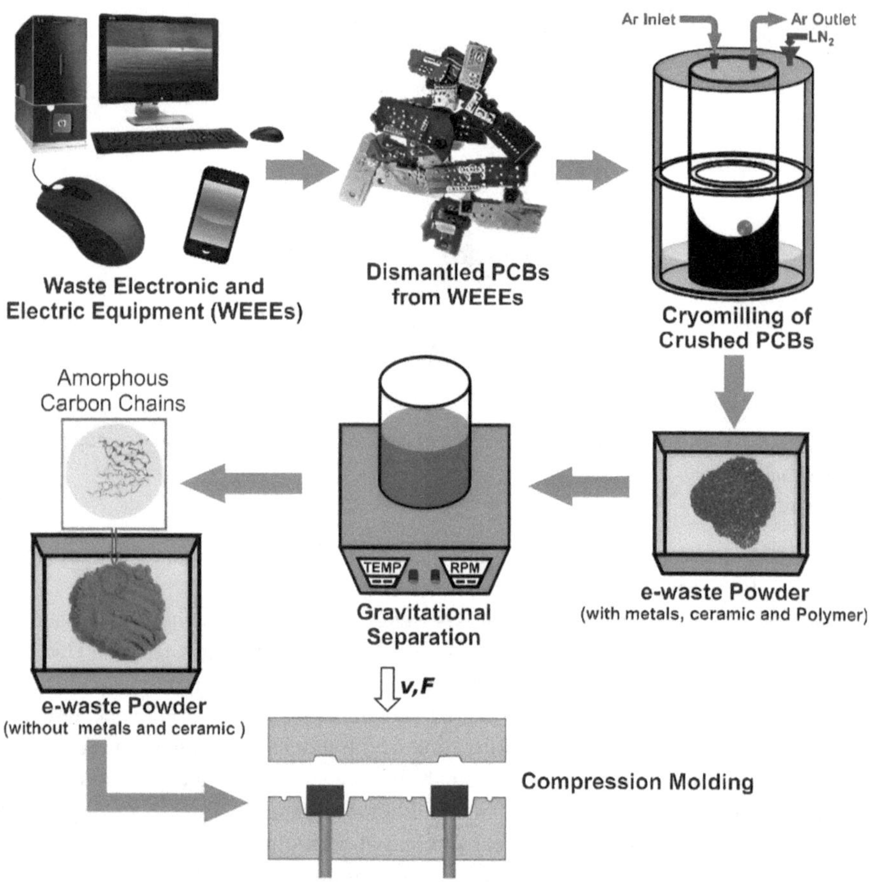

FIGURE 6.6 Green route to recycle and use polymeric content (with some ceramic particles from E-waste) of E-waste for manufacture of PCB switch board.

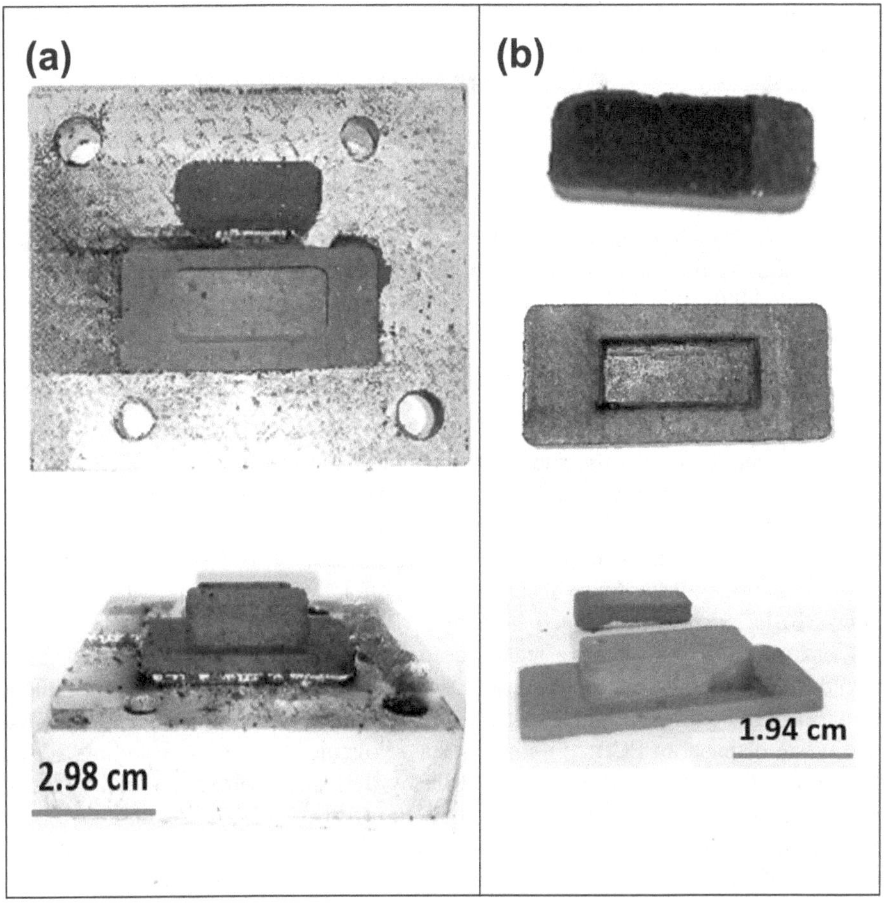

FIGURE 6.6 (Continued)

significance of the innovation goes beyond the technique of separation and extraction of polymers from waste PCBs. The reclaimed powdered polymeric serves as a significant resource and can also be utilized in the industrial sector. The idea presented herein focuses on the application of reclaimed polymeric E-waste powder in the production of electrical wall switch casings using compression molding techniques. A duralumin mold is fabricated for the purpose of producing a switch via the molding technique. Duralumin is chosen as a material for the mold due to its favorable thermal conductivity properties. Silicone oil is used as the lubricant to promote easy removal of fabricated components. Compression molding was performed at the optimum temperature, curing time and pressure. Subsequently, the mold, containing the fabricated electrical wall switch casing, was subjected to heat treatment in a furnace at a temperature of 280°C for a period of 2 hours. The concluding step involves the mechanical ejection of the formed switch casing from the mold. Figure 6.6 shows the molded components of the electrical wall switch casing using polymeric powder from waste

PCBs and its compression molding. This idea exemplifies the pragmatic efficacy of the concept since it furnishes a palpable utilization for the reclaimed materials. This incorporation of reclaimed polymeric E-waste in manufacturing processes showcases a pragmatic and economically feasible implementation of the innovation.

6.8 HOLISTIC MATERIALS CYCLE WITH E-WASTE RECYCLING

The E-waste, which forms a major chunk of waste in modern times, needs special attention and we need to move toward holistic materials cycle so that we can recover and recycle all the constituents with zero waste. This will make the processing route sustainable. By using innovation involving the novel amalgamation of mechanical separation, cryo-temperature milling, density-based separation, and the utilization of reclaimed materials for use.

The printed circuit board (PCB) mounted with electronic components (ECs) is the center-piece of all electric and electronic equipment (EEE). Waste printed circuit boards (WPCBs) account for 6% of the total weight of E-waste and are resource-rich but hazardous to human health and the environment [15]. These WPCBs contain a substantial amount of valuable metals (e.g., Cu, Fe, Al, Au, Ag, and Pd) but also contain many hazardous substances (e.g., Pb, Cr, Cd, polybrominated diphenyl ethers (PBDEs), polybrominated biphenyls (PBBs) [16–17]. Sensing the urgency and risk, several cost-effective techniques, mainly targeting metal recovery, have been developed. However, these conventional recycling technologies are not without limitations. In contrast, recent investigations by authors group [19–20] demonstrated zero waste disposal of PCB components via cryogenic temperature grinding, which has attracted significant scientific attention by the near-net recovery of individual metals for their contemporary utilization. Such investigations act as a directive light to guide the path of complete and green recycling of E-waste. However, a significant part of the PCBs, i.e., the polymeric component, remains unused and can be tapped into as the precursor for developing new materials and future technologies is an exciting subject worth exploration.

REFERENCES

[1] Jha R, Mishra G, Agrawal M, et al. Opportunities for an en-route to polymer inclusion membrane approach from conventional hydrometallurgical recycling of WPCBs: A mini-review. *Can Metall Q* 2022; 0: 1–15.

[2] Jha R, Rao MD, Meshram A, et al. Potential of polymer inclusion membrane process for selective recovery of metal values from waste printed circuit boards : A review. *J Clean Prod* 2020; 265: 121621.

[3] Trivedi H, Meshram A, Gupta R. Recycling of photovoltaic modules for recovery and repurposing of materials. *J Environ Chem Eng* 2023; 11: 109501.

[4] Kumar H, Meshram A, Gupta R. Removal of encapsulant Ethylene-vinyl acetate (EVA) from solar cells of photovoltaic modules (PVMs). *Mater Today Proc* 2023: 112, 73-76. DOI: 10.1016/j.matpr.2023.08.109

[5] Mishra G, Jha R, Meshram A, et al. A review on recycling of lithium-ion batteries to recover critical metals. *J Environ Chem Eng* 2022; 10: 108534.

[6] Verma HR, Singh KK, Mankhand TR. Liberation of metal clads of waste printed circuit boards by removal of halogenated epoxy resin substrate using dimethylacetamide. *Waste Manag* 2017; 60: 652–659.
[7] Verma HR, Singh KK, Mankhand TR. Comparative study of printed circuit board recycling by cracking of internal layers using organic solvents- dimethylformamide and dimethylacetamide. *J Clean Prod* 2017; 142: 1–7.
[8] Mishra G, Jha R, Rao MD, et al. Recovery of silver from waste printed circuit boards (WPCBs) through hydrometallurgical route: A review. *Environ Challenges* 2021; 4, 100073.
[9] Mishra G, Jha R, Meshram A, et al. A review on recycling of lithium-ion batteries to recover critical metals. *J Environ Chem Eng* 2022; 10: 108534.
[10] Azimi G, Sauber ME, Zhang J. Technoeconomic analysis of supercritical fluid extraction process for recycling rare earth elements from neodymium iron boron magnets and fluorescent lamp phosphors. *J Clean Prod* 2023; 422: 138526.
[11] Verma HR, Singh KK, Mankhand TR. Comparative study of printed circuit board recycling by cracking of internal layers using organic solvents-dimethylformamide and dimethylacetamide. *J Clean Prod* 2016; 142: 1–7.
[12] Verma HR, Singh KK, Mankhand TR. Recycling of waste printed circuit boards by delamination using different organic solvents and study of its mechanism. *254th ACS Meet Expo 20–24 Aug' 2017* 2017; 65: 139–146.
[13] Verma HR, Singh KK, Mankhand TR. Liberation of metal clads of waste printed circuit boards by removal of halogenated epoxy resin substrate using dimethylacetamide. *Waste Manag* 2016; 60: 652–659. DOI: 10.1016/j.wasman.2016.12.031
[14] Perkins DN, Brune Drisse MN, Nxele T, Sly PD. E-waste: A global hazard. *Annals Global Health* 2014: 80: 286–295.
[15] Lepawsky J. The changing geography of global trade in electronic discards: Time to rethink the e-waste problem. *Geogr J* 2015; 181(2): 147–159.
[16] Borthakur A, Govind M. Emerging trends in consumers' E-waste disposal behaviour and awareness: A worldwide overview with special focus on India. *Resour Conserv Recycl* 2017; 117: 102–113.
[17] Biswas S, Schoeberl A, Hao Y, et al. Integrated multilayer stretchable printed circuit boards paving the way for deformable active matrix. *Nat Commun* 2019; 10(1): 1–8.
[18] Chen Z, Yang M, Shi Q, Kuang X, Qi HJ, Wang T. Recycling waste circuit board efficiently and environmentally friendly through small-molecule assisted dissolution. *Sci Rep* 2019; 9(1): 1–9.
[19] Sharma N, Bajpai A, Yadav PK, Nellaiappan S, Sharma S, Tiwary CS, Biswas K. Green route for beneficiation of metallic materials from electronic waste for selective reduction of CO_2. *ACS Sustain Chem Eng* 2020; 8(32): 12142–12150.
[20] Bajpai A, Kumbhakar P, Tiwary CS, Biswas K, Conducting graphene synthesis from electronic waste. *ACS Sustain Chem Eng* 2021; 9(42): 14090–14100.
[21] Kumar N, Biswas K. Fabrication of novel cryomill for synthesis of high purity metallic nanoparticles. *Rev Sci Instruments* 2015; 86: 083903.

7 Emerging Sustainable Waste Management Technologies

Anurag Bajpai

7.1 INTRODUCTION

In the contemporary era characterised by rapid technological advancements, our global society is distinguished by remarkable levels of innovation and forward momentum. Nevertheless, the swift progression of technology also engenders a sometimes disregarded paralleled outcome: the generation of technological waste. This waste generated by our modern activities, ranging from the residues of metal extraction and mining operations to the discarded electronic gadgets that have grown pervasive in our daily lives, is a significant and daunting challenge (Aznar-Sánchez et al. 2018; Kiddee, Naidu, and Wong 2013). In addition to the issue of electronic waste, society is faced with the ongoing challenge of plastic pollution, the potentially harmful chemical by-products generated by industrial activities, and the formidable concern around the existence of radioactive waste (Evode et al. 2021; Misra and Pandey 2005). Nevertheless, as discussed throughout this book, these waste streams also pose an opportunity for us to utilise them as secondary resources to develop new materials and technologies. However, the prevalent practices relating to the concept of technological waste management and beneficiation are often intrusive to the environment and, therefore, cannot be considered sustainable (Rautela et al. 2021; Kalisz et al. 2022). The global landscape is transforming, necessitating a corresponding adaptation in our waste management strategies. By exploring scientific principles and unveiling new sustainable and green technologies, humanity can aim to progress towards the realisation of a cleaner, more environmentally sustainable, and peaceful global society.

The present chapter sheds light on the cutting-edge and ecologically conscious methodologies transforming the technological waste management paradigm. These strategies are not being adopted as temporary fixes for an escalating problem but as catalysts for fostering a more enduring and environmentally responsible future. The discussion involves a comprehensive analysis of the fundamental notion of sustainable waste management and the fundamental concepts that serve as the foundation of this crucial subject. Building upon the theoretical framework, the chapter will transition to a comprehensive examination of sophisticated methodologies for managing technological waste. The extraction of metals and mining waste, electronic waste, plastic waste, chemical waste, and nuclear waste each require specific procedures,

DOI: 10.1201/9781003619680-7

necessitating an examination of the scientific concepts that underlie these strategies. The examination of each waste stream is conducted in isolation, followed by an examination of the potential integration of various processes to provide comprehensive solutions.

7.2 THE CONCEPT OF SUSTAINABLE WASTE MANAGEMENT

Sustainable waste management is a comprehensive strategy that incorporates scientific principles, environmental stewardship, social equality, and economic viability to tackle the intricate challenges associated with waste generation and disposal (Agamuthu, Khidzir, and Hamid 2009). The overarching objective of such a comprehensive idea must aim to convert waste from a liability into a profitable asset, taking into account the complex interplay among natural systems, human communities, and economic systems. The ensuing principles of sustainable waste management are broadly discussed here.

- **Waste minimisation and prevention** are fundamentally grounded in the ideas derived from the disciplines of materials engineering, product design, and industrial ecology. Scientific research plays a crucial role in informing the advancement of environmentally sustainable materials and industrial methods with the objective of minimising waste generation from its origin. Methods such as material flow analysis and life cycle assessment (LCA) offer quantitative approaches for assessing the environmental impact of items and activities, aiding waste minimisation (Driussi and Jansz 2006).
- Scientific breakthroughs in materials characterisation, chemistry, and mechanical engineering are crucial in optimising **recycling processes and resource recovery**. Sophisticated sorting techniques utilise spectroscopy, X-ray fluorescence, and artificial intelligence to effectively segregate items (Bernat 2023). These systems use several sensors, including infrared spectroscopy and hyperspectral imaging, in order to accurately detect and separate items. The integration of robotics and automation systems significantly enhances the efficiency of these sorting procedures (Luo et al. 2022). Simultaneously, ongoing research in metallurgy, polymer science, and chemistry is dedicated to investigating novel approaches to **recovering valuable materials** from waste streams.
- The concept of **waste valorisation and upcycling** involves the scientific endeavour of optimising the extraction of value from waste materials. This entails the utilisation of novel methodologies to convert waste into products, chemicals, or materials of significant worth (Nyakuma and Ivase 2021). Thorough investigation in disciplines such as chemical engineering, biotechnology, and materials science plays a crucial role in the advancement of valorisation techniques. For example, the utilisation of waste plastics as feedstock for the production of 3D printing materials exemplifies advanced applications of waste valorisation (Mikula et al. 2021).

- **Waste-to-energy** systems leverage principles from thermodynamics, combustion science, and fluid dynamics to transform waste materials into energy that may be effectively utilised. Ongoing scientific investigations in the fields of catalysis, microbiology, and chemical engineering persistently develop these processes to achieve increased yields, minimised environmental consequences, and improved resource recuperation (Kothari, Tyagi, and Pathak 2010). Researchers consistently improve these procedures in order to optimise energy extraction while simultaneously reducing emissions and mitigating environmental effects.
- The field of **data analytics and monitoring** encompasses the systematic analysis and observation of data to extract meaningful insights and track various metrics. The use of data science, machine learning, and optimisation approaches is crucial in the realm of waste management operations. These technologies allow predictive analytics in waste collection, route optimisation, and demand forecasting, resulting in enhanced operational efficiency, cost reduction, and resource allocation optimisation (Diez-Olivan et al. 2019).
- The **Intersection of Equity and Environmental Justice** through the utilisation of geographic information systems (GIS), spatial epidemiology, and sociodemographic data in scientific investigations plays a crucial role in the identification and resolution of inequities pertaining to waste facility siting, access to recycling programmes, and the distribution of environmental burdens. The concept of social sustainability is intricately connected to the notion of environmental justice, necessitating the conduct of empirical research and fostering multidisciplinary cooperation (O'Neill Marie et al. 2003).
- The achievement of **economic sustainability** in waste management is contingent upon the effective usage of recovered resources in the field of resource economics. Researchers specialising in resource economics evaluate the economic viability of recycling by considering factors such as market dynamics, pricing mechanisms, and externalities. Cost-benefit and cost-effectiveness assessments provide decision-makers with valuable insights into the economic feasibility of waste management solutions.
- The notion of a **circular economy** provides a fundamental scientific basis for implementing sustainable waste management practices. Within the framework of a circular economy, waste is perceived as a valuable resource with the potential to be perpetually reintegrated into the production cycle. The exploration of this notion necessitates comprehensive investigation in disciplines such as systems thinking, industrial ecology, and materials science. In the circular economy framework, systems thinking is a conceptual framework that focuses on understanding the interconnections and interdependencies between various components within a system. It involves analysing and recognising the interdependencies among different elements within the waste management system. Further, the field of materials science plays a pivotal role in the attainment of circularity. Scientists are now engaged in developing materials that exhibit enhanced durability, recyclability, and environmental sustainability. Recent scientific advancements are significantly influencing

the field of materials science in several domains, such as bioplastics, self-healing materials, and nanotechnology.

7.3 ADVANCED TECHNOLOGICAL WASTE MANAGEMENT TECHNIQUES

While all the principles discussed in the preceding discourse hold significant value in the context of technological waste management and best practices, as individuals of scientific temper, we must delve deeper into the ideas and technologies that work on the principles of resource recovery and waste valorisation of technological waste. This section explores the theoretical principles and implementations of novel approaches to address distinct streams of technological waste. These technologies are being practised in several industrial sectors to transform the problem of technological waste into an opportunity.

7.3.1 Metal Extraction and Mining Waste

The process of metal extraction and mining operations produces substantial quantities of waste materials, such as tailings, slag, and smelting leftovers. These waste materials frequently consist of dangerous compounds, such as heavy metals, which have the potential to cause pollution of land and water (Krishnan et al. 2021). The implementation of sustainable management practices is crucial in order to effectively address environmental degradation and safeguard important resources, such as metals and minerals derived from these waste materials.

7.3.1.1 Sensor-Based Ore Sorting

The principle of sensor-based ore sorting is to use sensors to identify and separate valuable minerals from waste materials based on their spectral and physical properties. The efficacy of this approach is contingent upon integrating sophisticated sensors, data processing, and artificial intelligence algorithms to facilitate instantaneous decision-making in the sorting process (Robben and Wotruba 2019). The confluence of cutting-edge technology, mineral processing skills, and environmental aims is shown by sensor-based ore sorting. **Table 7.1** summarises the detection systems applied in mining operations (Robben and Wotruba 2019). The utilisation of sensor-based ore sorting technology is widely seen in mining operations as a means to optimise resource extraction and minimise wastage. It is also aimed at minimising the ecological footprint associated with mining activities. This is achieved by the reduction of waste creation and surface disturbances, hence mitigating the negative environmental consequences of mining operations (Modise et al. 2022).

The sensor-based sorting system developed by TOMRA Sorting Solutions had a notable influence on the operations of the Minsur San Rafael Tin Mine in Peru. The technique demonstrated a notable improvement in ore recovery rates, achieving a 10% increase through the precise identification and segregation of tin-bearing ore from waste rock (Robben and Wotruba 2019). Its implementation resulted in an

TABLE 7.1
Detection systems for sensor-based sorters and exemplary applications in mining

Detection Type	Material Property	Mineral Application
Radiometric	Natural gamma radiation	Uranium, precious metals
X-ray transmission	X-ray attenuation coefficient	Base/ precious metals, coal, diamonds, etc.
X-ray fluorescence	Elemental composition	Base/precious metals
X-ray luminescence	Visible luminescence under X-rays	Diamonds
Visual spectrometry	Reflection/absorption of visible radiation	Metals, industrial minerals, gem stones
Colour	Colour, reflection, brightness, transparency	Base/precious metals, industrial minerals, gem stones
Photometric	Monochromatic reflection, absorption	Industrial minerals, gem stones
Near-infrared spectrometry	Reflection/absorption of NIR radiation	Base metals, industrial minerals
Thermal infrared	Microwave excitation and thermal infrared detection	Base metals, precious metals
Laser triangulation	Hull detection (shape and form)	Base metals, precious metals, ferrous metals
Prompt gamma neutron activation analysis	Absorption and emission of prompt gamma rays	Ferrous metals
Laser induced breakdown spectrometry	Evaporation of matter and	Industrial minerals
Laser induced fluorescence	Absorption of laser light and spontaneous emission of light	Industrial minerals

Source: Robben and Wotruba 2019.

increase in the total production of valuable commodities while simultaneously reducing the adverse environmental effects caused by the mining operation.

7.3.1.2 Tailings Reprocessing

Tailings reprocessing encompasses the utilisation of sophisticated separation and hydrometallurgical techniques to retrieve precious metals or minerals from previously discarded mining tailings (Edraki et al. 2014). The reprocessing of tailings is a proactive strategy in the realm of mining waste management. The underlying theory revolves around the notion of harnessing untapped resources from preexisting waste streams. Tailings reprocessing has potential prospects in the following areas:

- **Resource Recovery:** The process of extracting precious metals from tailings to prolong mining activities' operational lifespan.
- **Waste Reduction:** The implementation of strategies aimed at minimising the quantity of legacy tailings, hence mitigating potential environmental hazards.

The reprocessing of previous gold mine tailings in South Africa by DRDGold serves as a noteworthy illustration of this technology. The implementation resulted in the retrieval of more than one million ounces of gold. Through the process of revisiting waste materials, DRDGold successfully converted them into valuable resources, therefore showcasing the economic feasibility and long-term profitability of tailings reprocessing (Durand 2012). **Figure 7.1** depicts a schematic representation of the reprocessing and use of medium and highly sulfidic mine tailings in the fabrication of construction materials (Martins et al. 2021).

Tailings reprocessing utilising the dry stack tailings method is another innovative strategy for the management of mining waste, which effectively curtails water use and mitigates adverse ecological consequences. The scientific principles in this context cover materials science, geotechnical engineering, and water conservation (Cacciuttolo and Valenzuela 2022). The methodology entails the utilisation of mechanical means to dehydrate tailings, therefore eliminating surplus water and yielding a compact, solid residue suitable for stacking. The utilisation of dry stack tailings is a viable solution to mitigate the necessity for water storage and management, hence resulting in a reduction in water consumption and the potential for pollution. The implementation of dry stack tailings at the Pueblo Viejo mine was undertaken to reduce water use and optimise their reprocessing. The use of this strategy resulted in a notable reduction in the environmental impact by effectively removing the requirement for expansive tailings ponds (Marsden and Botz 2017). This demonstrates the transformative potential of applying scientific ideas from the field of materials science and engineering to the domain of mining waste management.

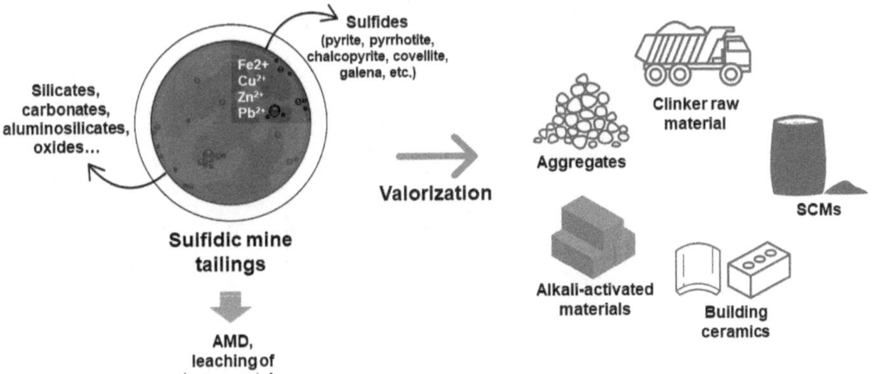

FIGURE 7.1 Schematic of utilisation of medium and highly sulfidic mine tailings in construction materials. (Martins et al. 2021.)

7.3.1.3 Bioleaching

The process of bioleaching utilises microorganisms to extract metals from ores of low-grade and mining waste materials. This methodology leverages the metabolic processes of microbes, enabling them to preferentially interact with particular metals and enhance their solubility (Baniasadi et al. 2019). The process of bioleaching exemplifies the integration of biological principles with metallurgical techniques in order to accomplish the extraction of valuable resources while mitigating any environmental damage. It serves as a prime example of sustainable mining operations that utilise nature's powers. Microbial activity, the selective affinity of microbes towards certain metals, facilitating their solubilisation and reduced energy consumption are three scientific concepts that underscore this technology (Borja et al. 2016). Bioleaching has several uses, one of which is in processing low-grade ores. This procedure aims to improve the extraction of metals such as copper, zinc, and gold from ores that are often difficult to process using conventional methods (Nkuna et al. 2022). It also assists in mitigating the environmental consequences associated with mining waste by the extraction of metals from abandoned sites and provides an environmentally friendly alternative to traditional hydrometallurgical techniques. **Figure 7.2** shows the biohydrometallurgy-based bioleaching technique as a promising method for the purpose of extracting metals from iron and sulfide ores, as well as minerals (Sajjad et al. 2019).

The utilisation of bioleaching at the Olympic Dam copper-uranium mine in Australia is an excellent method for economically extracting copper from ores with low-grade concentrations (Giurco et al. 2014). Bioleaching is also employed at the Bingham Canyon Mine located in Utah for the purpose of extracting copper from low-grade ore and tailings. This sustainable technology effectively mitigates waste generation and minimises resource consumption while exemplifying the scientific

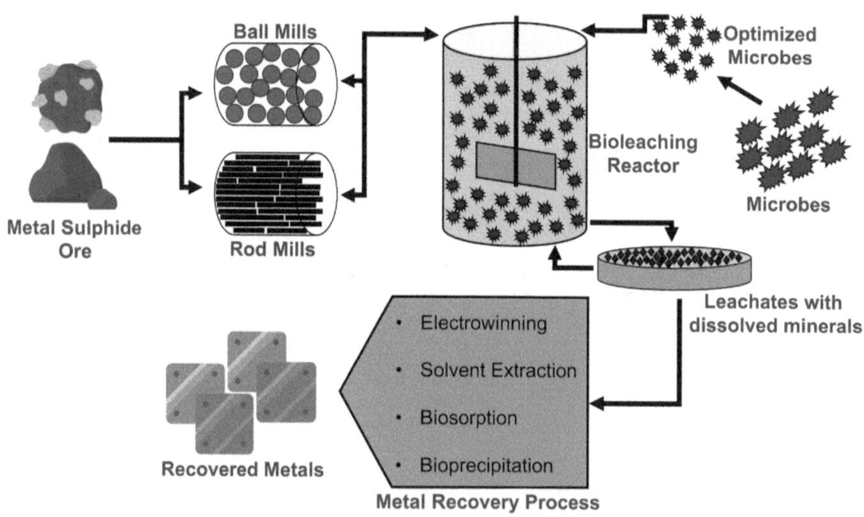

FIGURE 7.2 Flow sheet of bioleaching for low-grade sulfide ores. (Sajjad et al. 2019.)

implementation of microbiological and metallurgical principles in the realm of mining waste management (Olson, Brierley, and Brierley 2003).

7.3.1.4 Geopolymerisation

The process of geopolymerisation involves the conversion of mining waste materials into geopolymers, which are characterised by their exceptional structural properties. These geopolymers are inorganic polymers that exhibit notable performance attributes (Mohajerani et al. 2019). The geopolymerisation process represents a prime example of the harmonious integration of material science principles with sustainable waste management practices. The method is predicated upon the chemical reactions that ensue when alumino-silicate substances interact with alkaline solutions. Geopolymerisation has a wide range of uses, encompassing:

- **Waste Minimisation**: The implementation of strategies aimed at reducing the quantity of mining waste materials by their conversion into geopolymers.
- **Development of Sustainable Construction Materials**: The exploration of alternative construction materials with reduced environmental impact, including the development of concrete alternatives.

The utilisation of geopolymer concrete, derived from mining waste materials, is increasingly being adopted in construction endeavours inside the Philippines (Doiphode, Dhanuskar, and Deshmukh). The use of this sustainable substitute for conventional concrete showcases its capacity to mitigate the environmental consequences associated with both mining waste disposal and construction activities.

7.3.1.5 Subsurface Sequestration of Tailings

The subsurface sequestration of tailings is another novel strategy for the management of mining waste, whereby the disposal of tailings underground is employed to mitigate surface disruptions and adverse environmental effects. The subsurface sequestration of tailings involves the utilisation of geological, hydrogeological, and technical principles to provide secure and dependable containment strategies for mining waste (Khalidy and Santos 2021). However, subsurface disposal necessitates meticulous evaluation and choice of geological formations that can securely confine tailings and impede their dispersion into the surrounding environment. Further, comprehending the hydrogeological characteristics of the disposal site is crucial to mitigate the risk of water penetration and the subsequent leaching of toxins. Besides, the consideration of long-term stability and durability is important to the design and construction of subsurface disposal facilities, as it aims to guarantee the isolation of tailings.

The use of underground tailings disposal at the Olympic Dam mine in Australia aims to mitigate surface disruptions and minimise environmental damage. The technology entails the deposition of tailings into subterranean holes located within geologically appropriate formations (Cacciuttolo and Marinovic 2023). This case study demonstrates the application of scientific ideas in the fields of geology, hydrogeology, and engineering to enhance the sustainability of tailings management practices.

7.3.1.6 Hydrometallurgical Recycling of Mining Waste

The process of hydrometallurgical recycling of mining waste involves the extraction of precious metals from smelting slag and wastes through the utilisation of chemical procedures. Hydrometallurgical procedures utilise leaching processes in order to selectively dissolve certain metals from waste sources. Chemical reactions are then employed to precipitate and extract metals from leach solutions. The resource recovery process, namely through hydrometallurgical recycling, serves as a means to preserve valuable metals by recovering them from various waste streams (Matinde, Simate, and Ndlovu 2018).

The Kwinana nickel refinery, located in Western Australia, employs hydrometallurgical techniques with the purpose of extracting precious metals from smelting slag and waste. The use of this sustainable methodology effectively mitigates waste generation and minimises the environmental footprint through the efficient extraction and recycling of metals (Faris et al. 2023). This case study demonstrates the practical use of chemical and metallurgical concepts in the field of mining waste management.

7.3.1.7 Remote Sensing and GIS for Monitoring

The utilisation of remote sensing and geographic information system (GIS) technology is employed in the surveillance of mining waste, including monitoring tailings dams and establishing early warning systems. The utilisation of remote sensing and GIS technology places significant emphasis on the application of scientific concepts to improve the monitoring and safety of mining waste facilities (Song et al. 2020). In this strategy, satellite photography plays a crucial role in remote sensing, as it enables the collection of data pertaining to mining waste facilities and their respective surroundings through specialised sensors. The integration of data in GIS technology facilitates the thorough analysis and monitoring of diverse data sources. Early warning systems are technological tools that aid in the timely identification of possible environmental problems and dangers that may arise from the disposal of mining waste (Park and Choi 2020).

The Bulyanhulu Gold Mine in Tanzania employs remote sensing and GIS technologies to enhance the monitoring of the tailings dam and implement early warning systems. These technologies offer instantaneous data on the stability of dams, water levels, and potential environmental consequences, therefore augmenting safety measures and environmental management practices (Sillitoe and Thompson 2005). This exemplification showcases the successful utilisation of scientific ideas pertaining to remote sensing and geospatial analysis for the purpose of monitoring mining waste facilities.

7.3.2 Electronic Waste

As previously elaborated, electronic waste is composed of abandoned electronic gadgets that contain potentially harmful substances such as heavy metals, flame retardants, and toxic compounds. Insufficient management practices may lead to the occurrence of environmental contamination and pose potential health hazards

(Rautela et al. 2021). The use of sustainable practices in managing electronic waste is of utmost importance to address environmental pollution and facilitate the recovery of precious resources, such as metals and rare earth elements. One of the fundamental scientific ideas is Design for Disassembly (DfD), which involves the intentional design of products to facilitate their disassembly and promote the reuse of their components (Soh, Ong, and Nee 2014). Other innovative strategies include utilising cutting-edge approaches to effectively disassemble and recycle electronic items and effectively addressing and overseeing the handling and management of hazardous compounds found in electronic waste. Such strategies are discussed here.

7.3.2.1 Urban Mining and Automated Disassembly

Urban mining, sometimes referred to as "above-ground mining" or "circular mining," is an environmentally conscious strategy for resource management that, in the context of electronic waste, entails the retrieval of precious metals and minerals from electronic waste. The scientific foundations that form the basis of urban mining cover several disciplines, including materials science, resource economics, and environmental stewardship (Murthy and Ramakrishna 2022). The initial step in urban mining is conducting a comprehensive examination of the composition of elements found in electronic waste. This utilises sophisticated methodologies, namely X-ray fluorescence (XRF) spectroscopy and scanning electron microscopy (SEM), to discern and measure the elemental and mineralogical constituents. Further, metallurgical techniques are then utilised for the extraction of precious materials once they have been discovered. Methods such as hydrometallurgy and pyrometallurgy have been employed to selectively extract metals such as gold, silver, copper, and rare earth elements from electronic waste (Xavier, Ottoni, and Abreu 2023; Xavier et al. 2021).

In order to facilitate comprehension of the legal, economic, and geographic features of urban mining of electronic waste, **Table 7.2** presents an overview of case studies from various nations and locations. In addition, the process of extracting gold from electronic waste, specifically outdated computers and mobile phones, is a popular and prominent use case. Urban mining businesses utilise several sophisticated methodologies to extract gold from electronic waste. One example of a technique employed in the recovery of gold from intricate electronic components is known as "electrolytic gold recovery," which involves the application of electricity to selectively deposit gold (Sum 1991). Another innovative approach involves the process of recovering rare earth elements (REEs) from electronic waste for subsequent usage in electronic devices. This phenomenon mitigates the necessity for mining operations that have detrimental effects on the environment. Scientists have devised techniques for extracting REEs from electronic gadgets that have been abandoned, therefore mitigating the environmental impact associated with the extraction of REEs (Ramprasad et al. 2022). The recovery of copper from wiring is another process that involves extracting the extremely expensive metal from electrical wiring and other electronic components. The primary objective of urban mining procedures is to achieve optimal copper recovery by employing methods such as electrochemical deposition. This technology allows the selective plating of copper from electronic waste materials (Abdelbasir et al. 2018).

TABLE 7.2
Aerial view of case studies on e-waste urban mining

Region /Country	Characteristics	E-waste (kg per capita in 2019)	Sources
Africa	Informal collectors and recyclers dominate e-waste management, which lacks established take-back systems and licensing requirements for sorting and disassembling e-waste. Only a few nations have issued e-waste laws (e.g., Egypt, Madagascar, Ghana, Rwanda, Nigeria, Cameroon, South Africa and Côte d'Ivoire), even though their implementation is ineffective.	2.5	(Asante, Amoyaw-Osei, and Agusa 2019)
Americas	The Americas are one of the biggest e-waste generators worldwide. The greatest producers of e-waste in this continent are the higher-income areas (USA, Canada, Brazil and Mexico).	13.3	(Murthy and Ramakrishna 2022)
Asia	The increasing population in urban areas and urban landscapes require better infrastructure to manage the anthropogenic stocks. The countries in the South Asia area (China, Japan, India, Korea and Singapore) have recognised the need for effective e-waste management and enacted legislation to manage e-waste.	5.6	(Johansson et al. 2016)
Europe	The majority of e-waste is regulated by the WEEE Directive (2012/19/EU), which is in force in the European Union and in Norway. The EU member states are required to support the design and manufacture of EEEs that can be disassembled and recycled, besides ensuring ecologically friendly processing under the extended producer responsibility (EPR) principle, which allows manufacturers to assume responsibility for recycling their end-of-life items.	16.2	(Forti et al. 2020)
Oceania	Australia is the only country in the area to have enacted legislation specifically addressing e-waste management. Following Australia's lead, New Zealand and several other Pacific Island governments are implementing product stewardship frameworks. Many of the island nations in this region have broad environmental protection legislation that includes waste management rules (including e-waste).	16.1	(Murthy and Ramakrishna 2022)
USA	The Resource Conservation and Recovery Act (RCRA), laws for recycling Cathode Ray Tubes (CRTs) and the National Strategy for Electronics Stewardship framework have all been used in the USA to manage domestic e-waste.	21	(Xavier, Ottoni, and Lepawsky 2021)

Canada	There is a current flux of e-waste from Canada to the US, driven by market conditions. The e-waste ends up at three main destinations: treatment (refurbishment, remanufacturing and recycling), disposal and exportation. E-waste is imported to Canada from Australia, USA, China, Denmark, Japan and Poland, and exportation fluxes from Canada to other countries (Singapore and South Africa). The e-waste management is performed by extended producer responsibility (EPR) programs, regulated by the provinces.	20.2	(Tutton, Young, and Habib 2022)
Brazil	Brazil is the greatest e-waste generator in South America, with diverse e-waste flows, and pursues regulatory mechanisms already implemented to manage those flows under legal environmental penalties. The reverse logistics system (RLS) targets were established from 2021 to 2025 for e-waste collection and disposal of all streams by the Brazilian Policy on Solid Waste (BPSW). The RLS is the basis of the urban mining potential. Municipal and state regulations were enacted before the BPSW. Instead of usual EPR program, in Brazil was established the shared responsibility.	10.2	(Xavier, Ottoni, and Lepawsky 2021)
China	Chinese environmental legislation covers 14 types of e-waste. The country exports e-waste to Southeast Asia, Africa and other regions. E-waste processing occurs from formal and informal activities, together with high technological solutions.	7.2	(Lepawsky 2015)
India	High population density but low e-waste consumption per capita. India is the only country in Southern Asia with e-waste legislation. Despites the prevalence of informality, the regulation is focused in setting formal recycling facilities and Networks to be enhanced for wider and proper treatment in the urban mining ontexto.	2.4	(Sharma, Joshi, and Govindan 2021)

Sources: sante, Amoyaw-Osei, and Agusa 2019; Murthy and Ramakrishna 2022; Johansson et al. 2016; Forti et al. 2020; Xavier, Ottoni, and Lepawsky 2021; Tutton, Young, and Habib 2022; Lepawsky 2015; Sharma, Joshi, and Govindan 2021.

Automated disassembly and subsequent mining using robotic systems, a fourth-generation industrial technology, encompasses the utilisation of sophisticated automation and robotics to effectively dismantle electronic equipment, therefore retrieving valuable components and resources (Cui and Forssberg 2003). The field of robotic automation encompasses developing sophisticated robotic systems that can effectively discern, dismantle, and categorise electronic components and materials. These robotic systems utilise a combination of sensors, machine vision technology, and artificial intelligence algorithms in order to accurately identify and manipulate different components of devices (Li, Barwood, and Rahimifard 2018). Following the process of disassembly, the retrieved materials undergo a systematic sorting and processing procedure to facilitate their recycling. Selective extraction is employed to get valuable metals such as aluminium, cobalt, and copper. Robotic systems are renowned for their exceptional precision and rapidity, hence greatly augmenting the efficacy of electronic waste recycling procedures (Sarc et al. 2019).

The "*Liam*" robot developed by Apple exemplifies the technological advancements in urban mining using robotics. *Liam* possesses a set of numerous appendages and optical devices, enabling it to dismantle iPhones proficiently while effectively discerning and retrieving important resources like aluminium, cobalt, and rare earth elements (Ghimire and Ariya 2020). The robot demonstrates the utilisation of principles in robotics engineering and automation to address the issue of electronic waste recycling.

7.3.2.2 Rare Earth Element Recovery

The process of recovering rare earth elements (REEs) from electronic waste is a prominent and promising attempt to tackle both the growing electronic waste as well as the depleting REE deposits. This entails the extraction of these precious elements from discarded electronic equipment with the intention of reusing them in the production of new electronic devices. This practice contributes to the establishment of a more circular and environmentally conscious electronics sector. The approach is based on conserving resources and decreasing reliance on mining practices with negative environmental impacts. REE recovery is employed for the purpose of resource conservation, aiming to decrease the need for primary extraction of rare earth elements (REEs) that might potentially have significant environmental consequences (Ippolito et al. 2017; Zhi et al. 2022; Ramprasad et al. 2022). The various prevalent technologies and strategies for REE recovery from electronic waste have already been discussed to some length in previous chapters. **Table 7.3** presents a comprehensive overview of the extraction efficiencies of various REEs from a diverse array of electronic waste through the utilisation of multiple extraction techniques.

The Urban Mining Company, based in the Netherlands, focuses its efforts on extracting and retrieving rare earth metals from discarded fluorescent bulbs. By employing sophisticated metallurgical techniques, the extraction and purification of rare earth elements (REEs) are accomplished, therefore providing a viable and environmentally responsible supply of these crucial components for the production of electronic devices. This example showcases the utilisation of metallurgical concepts in recycling electronic waste.

TABLE 7.3
Recovery of different REEs from electronic waste using a variety of methods

Electronic Waste Stream	Method	REEs Extracted	% Recovery	Reference
Waste fluorescent powder	Hydrometallurgical method. Alkaline roasting-acid leaching process for the conversion of Al_2O_3 into water soluble $NaAlO_2$ so that it can be easily separated from REE oxides by filtration.	CeO_2, Eu_2O_3, Er_2O_3, Tb_4O_7, Y_2O_3,	99.8%	(Liao et al. 2017)
Lamp phosphor waste	Hydrometallurgical process based on H_2SO_4 leaching.	Ce, La and Tb	82% (Ce), 75% (La), and 82% (Tb)	(Yurramendi et al. 2019)
Waste cathode ray tube phosphor powder	Thermal pre-treatment integrated with the hydrometallurgical process. High temperature calcination followed by hydroxylation with NaOH to remove the PO_4^{3-}, Zn and other base metals.	Ce, Eu, Sm, and Y	58% (Y), 90% (Eu), 90% (Sm), and 87% (Ce)	(Alvarado-Hernández, Lapidus, and González 2019)
Fluorescent powder waste	Mechanical activation by a vibratory disc mill, followed by H_2SO_4 leaching.	Tb	98.3%	(Ippolito et al. 2021)
Discarded NdFeB magnets	Hydrometallurgical method. Carbonisation/hydrogenation-hydrolysis using biochar as an extractant.	Nd, Gd, Pr	88.4%	(Liu et al. 2019)
Fluorescent lamp powder	Sequential digestion of YOX and LAP, followed by separately treating the leachates under specific hydrometallurgical conditions.	Ce, Eu, La, Tb, and Y	>95%	(Patil et al. 2021)
Fluorescent lamp wastes	Hydrometallurgical method. Solid-state chlorination using dry HCl(g), converting the REE content of the Hg-free phosphor waste into water soluble REE metal chlorides.	Eu and Y	95.7% (Y) and 92.2% (Eu)	(Pavón et al. 2021)

(continued)

TABLE 7.3 (Continued)
Recovery of different REEs from electronic waste using a variety of methods

Electronic Waste Stream	Method	REEs Extracted	% Recovery	Reference
Nickel-metal hydride (NiMH) battery waste	Hydrometallurgical method. An ionic liquid [trihexyl(tetradecyl) phosphonium]$_2$[2,2′ -(1,2-phenylenebis(oxy)) dioctanoate] ([P6,6,6,14]$_2$[OPBOA]). The extracted Ni(II) and Co(II) was scrubbed and stripped by dil. NaCl solution. The loaded Nd(III) in [P6,6,6,14]2[OPBOA] was by dil. HCl, dil. $Na_2C_2O_4$ or dil. Na_2CO_3.	Nd	>96.4% and 99.8 wt% by NaCl and Na_2CO_3 stripping, respectively	(Su, Xie, and Sun 2021)
Waste cathode ray tube phosphors	Hydrometallurgical method. Oxidative leaching by H_2SO_4/H_2O_2 mixture.	Eu, Y	99% (Eu) and 87% (Y)	(Yin et al. 2018)
NdFeB battery waste	Hydrometallurgical method. H_3PO_4 leaching.	Nd, Pr, and Dy	98.76%	(Liu et al. 2021)
Green phosphors from fluorescent lamp waste	Thermal pre-treatment integrated with the hydrometallurgical process. Alkaline fusion with $Ba(OH)_2$ followed by H_2SO_4 leaching.	Y, Eu, Tb, La, Gd and Ce	99% (Y and Eu), 80% (Tb), 65% (La), 63% (Gd), and 60% (Ce)	(Ippolito et al. 2017)
NdFeB waste	Hydrometallurgical method. $H_2C_2O_4$/HCl leaching.	Nd, Pr	≤93.89%	(Liu et al. 2021)
Waste permanent magnets	Membrane technology. Solvent extraction followed by filtration through a hollow fibre supported liquid membrane.	Nd, Dy, and Pr	63.13% (Nd), 15.21% (Dy), and 56.29% (Pr),	(Ni'am et al. 2020)
Waste permanent magnets	Thermal pre-treatment integrated with the hydrometallurgical process. High temperature roasting, followed by HCl/H_2SO_4 leaching.	Nd, Dy, and Pr	78% (Nd), 83% (Pr), and 76% (Dy)	(Ni'am et al. 2019)

Computer monitor scraps	Hydrometallurgical method. HCl/H_2SO_4 leaching.	Eu, Y	>97%	(Resende and Morais 2010)
Waste phosphors	Mechanical activation using a ball-mill, followed by HCl leaching.	Tb, La, Ce	~90%	(Tan, Deng, and Li 2017)
Waste nickel-metal hydride batteries	Hydrometallurgical method. Phosphate leaching followed by HCl stripping.	La, Ce, Nd, and Pr	97.84% (La), 100% (Ce), 100% (Nd), and 99.77% (Pr)	(Zhi et al. 2022)
Permanent magnet waste	Electrochemical method. Electrolysis was undertaken in molten $LiF\text{-}CaF_2$.	Nd, Pr	Nd (97.25%), Pr (99.1%)	(Yang et al. 2020)
Spent NiMH batteries	Biometallurgical method. Extraction used 5-ketogluconate, gluconate, and 2-ketogluconate.	Y, La, Ce, Pr, Nd, Sm, Gd, Er, Yb	Y (95%), La (75%), Ce (78%), Pr (50%), Nd (58%), Sm (15%), Gd (78%), Er (43%), Yb (38%)	(Rasoulnia et al. 2021)

Sources: Liao et al. 2017; Yurramendi et al. 2019; Alvarado-Hernández, Lapidus, and González 2019; Ippolito et al. 2021; Liu et al. 2019; Patil et al. 2021; Pavón et al. 2021; Su, Xie, and Sun 2021; Yin et al. 2018; Liu et al. 2021; Ippolito et al. 2017; Ni'am et al. 2020; Ni'am et al. 2019; Resende and Morais 2010; Tan, Deng, and Li 2017; Zhi et al. 2022; Yang et al. 2020; Rasoulnia et al. 2021.

7.3.2.3 E-Waste Collection Formalisation

The formalisation of e-waste collection is the establishment of structured and controlled procedures for the collecting and recycling of electronic waste. This is done with the aim of mitigating the risks associated with informal recycling techniques, which can be harmful. The formalisation of e-waste collection highlights the significance of implementing organised and controlled procedures in order to effectively tackle the issues associated with e-waste, safeguard human well-being, and encourage the retrieval of valuable resources (Sengupta et al. 2022). By minimising the health and environmental dangers associated with the informal handling of e-waste, formalisation measures aim to safeguard the well-being of individuals and the ecosystem, enhancing the efficiency of extracting valuable materials from electronic waste and ensuring adherence to environmental norms and standards in order to maintain legal conformity.

The formalisation of e-waste collection and recycling operations by the e-Waste Association of South Africa (eWASA) has been accomplished with success. This campaign aims to redirect electronic waste away from potentially dangerous informal recycling methods and instead encourages the adoption of appropriate procedures for managing e-waste (Mahore, Mahore, and Dhoble 2023).

7.3.2.4 Closed-Loop Supply Chains: Reusing Components in New Products

The primary objective of closed-loop supply chains in electronic waste management is to prioritise the reuse of components derived from obsolete electronic devices to manufacture new goods. Closed-loop supply chains are an environmentally conscious strategy that uses scientific concepts to reduce resource utilisation and waste production within the electronics sector (Islam and Huda 2018). The scientific ideas encompassed in this context pertain to the fields of materials science, product design, and resource economics.

- The concept of **component reusability** is the intentional design of products for disassembly, wherein components are meticulously chosen and made with the purpose of being reused in future devices.
- The concept of **resource efficiency** refers to the optimisation and effective utilisation of resources in order to minimise waste and maximise output. Closed-loop supply chains effectively mitigate resource demand since they facilitate the extension of the operational lifespan of electronic components and materials.

The Closed-Loop Recycling Program implemented by Dell integrates recovered polymers derived from obsolete electronic devices into the production of new computer systems (Akçalı and Çetinkaya 2011). The closed-loop methodology encompasses the retrieval, manipulation, and reincorporation of plastic materials derived from discarded electronic equipment into the production of novel computer components. This study emphasises the scientific foundations underlying the practice of materials reuse and circular design within the context of electronic waste management.

7.3.3 Plastic Waste

Plastic waste encompasses a diverse array of polymers that exhibit long-lasting durability within the environment, enduring for extended periods of time, often spanning hundreds of years. Insufficient management of plastic waste results in the contamination of oceans and poses a threat to marine biodiversity (Evode et al. 2021). The use of sustainable practices in the management of plastic waste is crucial in mitigating environmental harm, curbing the proliferation of microplastic contamination, and facilitating the shift towards a circular economy for plastics. The fundamental scientific principles for developing suitable strategies to mitigate this waste encompass closed-loop recycling that effectively preserves the quality of materials throughout the recycling process, the development of biodegradable and recyclable materials that possess the properties of biodegradability and recyclability, as opposed to conventional plastics and plastic-to-fuel conversion, involving the transformation of plastic waste into usable fuels, hence mitigating waste accumulation.

7.3.3.1 Chemical Recycling of Plastics

The chemical recycling method entails employing sophisticated techniques to decompose polymers into their fundamental constituents, hence enabling the production of novel plastics or other useful materials (Thiounn and Smith 2020). The process of chemical recycling of plastics serves as a prime example of a profoundly inventive strategy in waste management, which effectively fosters the optimisation of resources and the preservation of environmental sustainability. Depolymerisation is the initial step in the process, wherein polymer chains are cleaved into smaller pieces. The introduction of heat, catalysts, or chemical agents is necessary, contingent upon the specific type of plastic involved. Chemical reactions are employed to transform depolymerised monomers into simpler molecules that are more suited for reuse (Rahimi and García 2017; Goto 2009). These reactions might include processes like hydrolysis or cracking. Product separation is a crucial step in the chemical synthesis process. After synthesising the necessary chemicals, several procedures, such as distillation, extraction, or chromatography, are employed to effectively isolate and purify the products. **Figure 7.3(a)** illustrates the reaction strategy for the degradation of polyethylene terephthalate (PET) in a supercritical methanol environment. **Figure 7.3(b)** shows Nylon 6 undergoing hydrolysis to yield ε-aminocaproic acid, which was subsequently subjected to cyclodehydration to form ε-caprolactam (Goto 2009).

Chemical recycling of plastics is also utilised for plastic regeneration, which involves the conversion of mixed or contaminated polymers into high-quality feedstock that may be used to create new plastics (Peng et al. 2022). Polystyrene is a versatile material with several applications. Agilyx, a company at the forefront of chemical recycling, utilises a method that transforms wasted polystyrene waste, which is frequently seen in packaging materials and disposable products, into styrene monomers. This process entails the utilisation of high-temperature pyrolysis, wherein polystyrene is subjected to heat in an oxygen-deprived environment, resulting in the depolymerisation of the material and the subsequent liberation of styrene monomers (Thiounn and Smith 2020). Additionally, polyethylene terephthalate (PET), a material often employed in producing beverage bottles and textile fibres, can potentially undergo

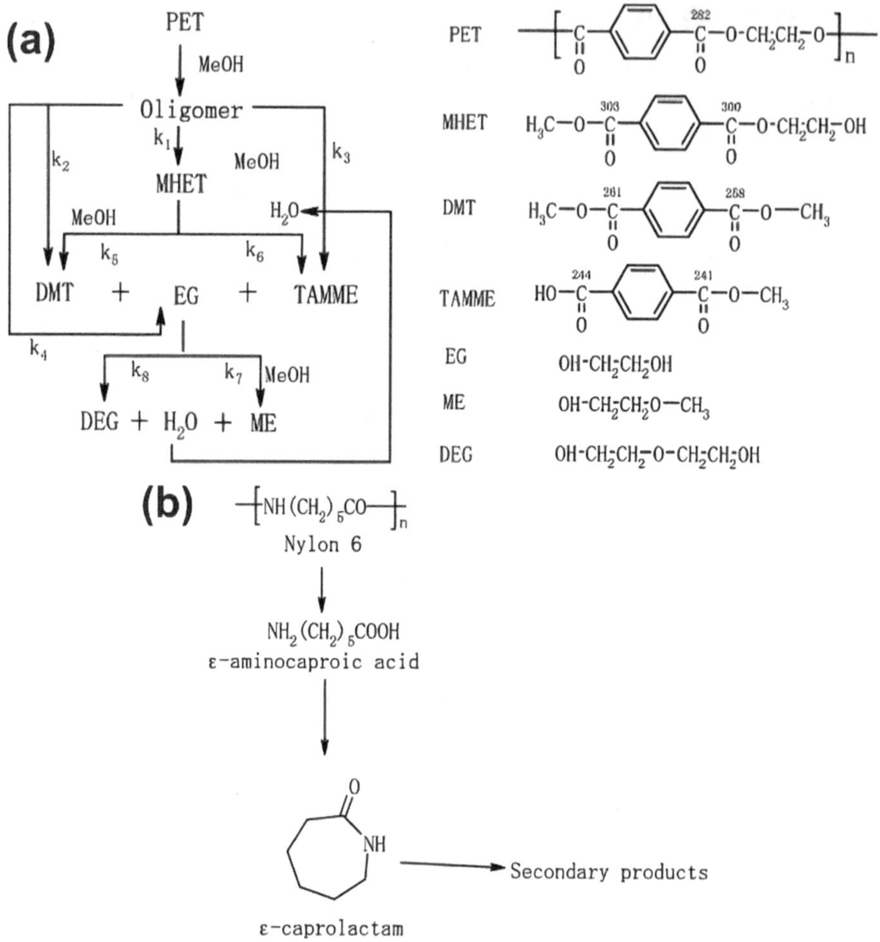

FIGURE 7.3 (a) Reaction scheme for decomposition of PET in methanol; (b) decomposition of Nylon 6 in subcritical water. (Goto 2009.)

chemical recycling processes that convert it into terephthalic acid. This acid is a crucial component in manufacturing fresh PET materials (Al-Sabagh et al. 2016). The aforementioned procedure frequently encompasses a synergistic use of depolymerisation, purification, and chemical processes in order to restore and renew vital feedstock. Nylon 6, a frequently encountered material in the textile and carpet industry, has the potential to undergo chemical recycling processes that convert it into caprolactam, a key intermediate compound utilised in the manufacturing of Nylon 6. The procedure involves the use of hydrolysis to disassemble the polymer, followed by several reaction stages to get caprolactam (Minor et al. 2023).

7.3.3.2 Mechanical Recycling With Advanced Sorting

The technique of mechanical recycling with improved sorting encompasses the application of state-of-the-art technologies to effectively separate and treat mixed plastics.

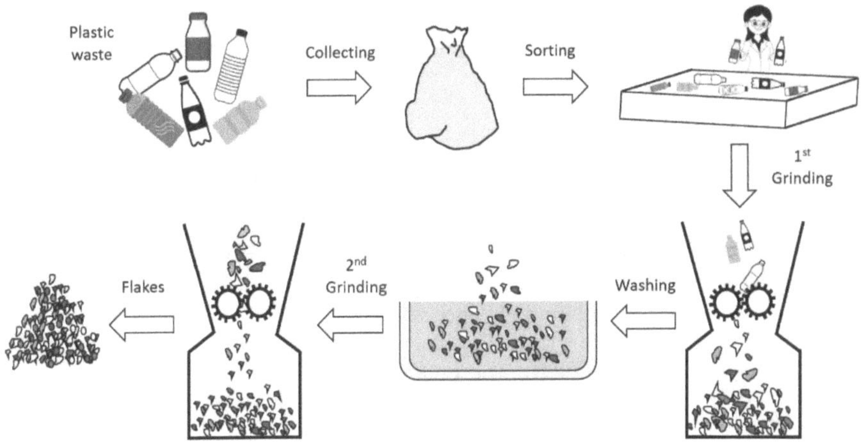

FIGURE 7.4 Scheme of the basic principal steps in a mechanical recycling process. (Kuzmanović et al. 2021.)

The utilisation of improved sorting techniques in mechanical recycling serves as a prime example of how contemporary technology and scientific ideas are integrated to optimise recycling procedures, minimise impurities, and bolster the overall sustainability of plastic recycling (Schyns and Shaver 2021). The mechanical recycling process encompasses many sequential stages, including collecting, sorting, washing, grinding (as seen in **Figure 7.4**), and subsequent reprocessing to fabricate novel recycled materials (Kuzmanović et al. 2021). The materials identification process involves using advanced sensing technologies, including near-infrared (NIR) spectroscopy and X-ray fluorescence (XRF), to discern and distinguish various types of plastics by analysing their unique spectral fingerprints or elemental makeup (Araujo-Andrade et al. 2021). Materials separation/sorting involves the utilisation of automated sorting equipment, such as conveyor belts equipped with air jets or robotic arms, to accurately divide plastic products into homogenous streams according to their specific polymer type. Quality control systems are used to oversee the sorting process, with the primary objective of guaranteeing the purity and quality of the separated materials. The diligent monitoring and assessment of the sorting process achieves this optimisation of recycling efficiency.

This technology plays a critical role in achieving efficient sorting, namely in the context of separating mixed polymers into homogenous streams for the purpose of recycling. It also ensures the preservation of material quality during recycling procedures. The AUTOSORT machine developed by TOMRA Sorting Solutions serves as a notable illustration of such cutting-edge sorting technology. The system effectively categorises heterogeneous plastics according to their respective polymer compositions using near-infrared (NIR) sensors and high-speed ejectors, hence improving recycling efficiency and mitigating contamination levels (Araujo-Andrade et al. 2021). The technology mentioned above is widely employed in recycling facilities throughout the globe with the aim of enhancing the calibre and efficacy of plastic recycling processes. In another example, the UniSort line of sorting machines developed by Steinert utilises X-ray transmission technology to discern and segregate various types of plastics based on their elemental composition. This method facilitates the segregation of visually identical polymers based on their distinct elemental characteristics.

7.3.3.3 Biodegradable Plastics

Biodegradable polymers are specifically engineered to undergo natural decomposition processes without causing any adverse effects on the environment. The core concept is around the advancement of polymers that can decompose into non-toxic compounds under specified environmental circumstances. Biodegradable plastics are engineered to possess distinct chemical features and architectures that enable their decomposition into natural chemicals when exposed to environmental conditions (Huang, Shetty, and Wang 1990). The composition of biodegradable plastics commonly consists of polymers that exhibit susceptibility to microbial assault and enzymatic breakdown. The polymers under consideration encompass polylactic acid (PLA), polyhydroxyalkanoates (PHA), and starch-based compounds. The breakdown process of biodegradable polymers is dependent on the presence and activity of microorganisms within the surrounding ecosystem. Microorganisms, such as bacteria and fungi, possess the ability to synthesise enzymes capable of catalysing the hydrolysis of polymer chains, resulting in the production of simpler molecular compounds (Ali et al. 2022). Environmental conditions, including temperature, humidity, and microbial activity, are significant elements that influence the process of biodegradation.

Biodegradable polymers find utility in a diverse range of applications, encompassing single-use products, such as disposable cutlery and packing materials, biodegradable mulch films in agricultural crop cultivation and biodegradable materials for specific medical uses in the field of medical devices. The biodegradable polymers produced by Novamont, known as Mater-Bi, serve as a notable illustration of accomplishment in this field. These materials find extensive usage across several domains, encompassing applications such as retail bags, packaging, and agricultural films. Mater-Bi polymers have been specifically engineered to undergo degradation within natural habitats, therefore mitigating the long-term presence of plastic waste and providing a viable and environmentally conscious substitute for conventional plastics (Bastioli 1998). In another case, polylactic acid (PLA) is a type of biodegradable plastic that is sourced from renewable materials, specifically maise starch. Polylactic acid (PLA)--based biodegradable polymers experience microbial breakdown within composting facilities, whereby carefully regulated environmental conditions expedite the process of disintegration into carbon dioxide and water. Polyhydroxyalkanoates (PHA) are a class of biopolymers that exist naturally and may be produced by microbial synthesis. Polyhydroxyalkanoate (PHA) biodegradable polymers can undergo degradation by microorganisms present in soil and water environments, hence playing a role in diminishing the long-lasting nature of plastic materials (La Fuente, Maniglia, and Tadini 2023).

7.3.3.4 Plastic-to-Fuel Conversion

The primary objective of plastic-to-fuel conversion technology is to transform polymers that cannot be recycled into fuels with high energy content, such as diesel or gasoline (Panda, Singh, and Mishra 2010). The conversion of plastic into fuel highlights the utilisation of thermochemical and catalytic principles in tackling the issue of non-recyclable plastic waste, presenting a viable and environmentally friendly approach to waste disposal and energy production. Pyrolysis is the fundamental technique employed in the conversion of plastic into fuel. The process involves subjecting plastic waste to heat in an oxygen-depleted environment, resulting in its decomposition into various forms, including gases, liquids, and solids. Some catalytic processes can be utilised to enhance the quality of liquid products derived from pyrolysis, transforming them into fuels of superior quality (Miandad et al. 2016). The ongoing efforts of researchers and engineers involve the development of catalytic methods for enhancing the quality of liquid products derived from plastic pyrolysis, aiming to produce fuels of superior quality. These methods entail the utilisation of catalysts to expedite chemical reactions, hence improving the characteristics of fuels. **Figure 7.5** illustrates the mechanism through which plastic is converted into fuels using photocatalytic oxidation, Fenton oxidation, electro-catalytic oxidation, and advanced oxidation (Li et al. 2022). Energy recovery is a crucial aspect of all the conversion processes since it yields significant energy in the form of liquid fuels. These fuels have many applications, such as powering transportation systems and generating electricity.

The Plastics to Oil (PTO) method developed by Blest Corporation utilises a pyrolysis process to transform plastic waste into a combination of gases and liquid hydrocarbons. These by-products can then undergo additional refining processes to produce various types of fuels. The utilisation of plastics as an energy source is exemplified by this technology, which serves to illustrate the practical implementation of thermochemical concepts within the realm of waste management.

7.3.4 Industrial Chemical and Radioactive Industrial Waste

The sustainable management of industrial chemical and radioactive industrial waste includes the mitigation of potential hazards to both human health and the environment. One of the fundamental scientific ideas is the concept of containment and isolation, which involves the secure confinement of hazardous waste materials in order to prevent any potential exposure or contamination (Li et al. 2012). Another is waste minimisation, involving the use of process innovations and recycling methods to effectively reduce the creation of waste. Long-term monitoring is the ongoing and systematic observation and evaluation of waste storage and disposal facilities.

The implementation of secure land disposal with numerous barriers is a crucial strategy for effectively handling hazardous chemical and radioactive industrial waste. The process entails the meticulous confinement and segregation of waste materials in order to avert pollution and safeguard the well-being of individuals and the ecosystem. The Multiple Barrier System is predicated upon utilising a complex network of barriers specifically engineered to effectively impede the dissemination of perilous chemicals. Common obstacles frequently seen in this context encompass many types of barriers, such as liners, artificial caps, and naturally occurring geological structures

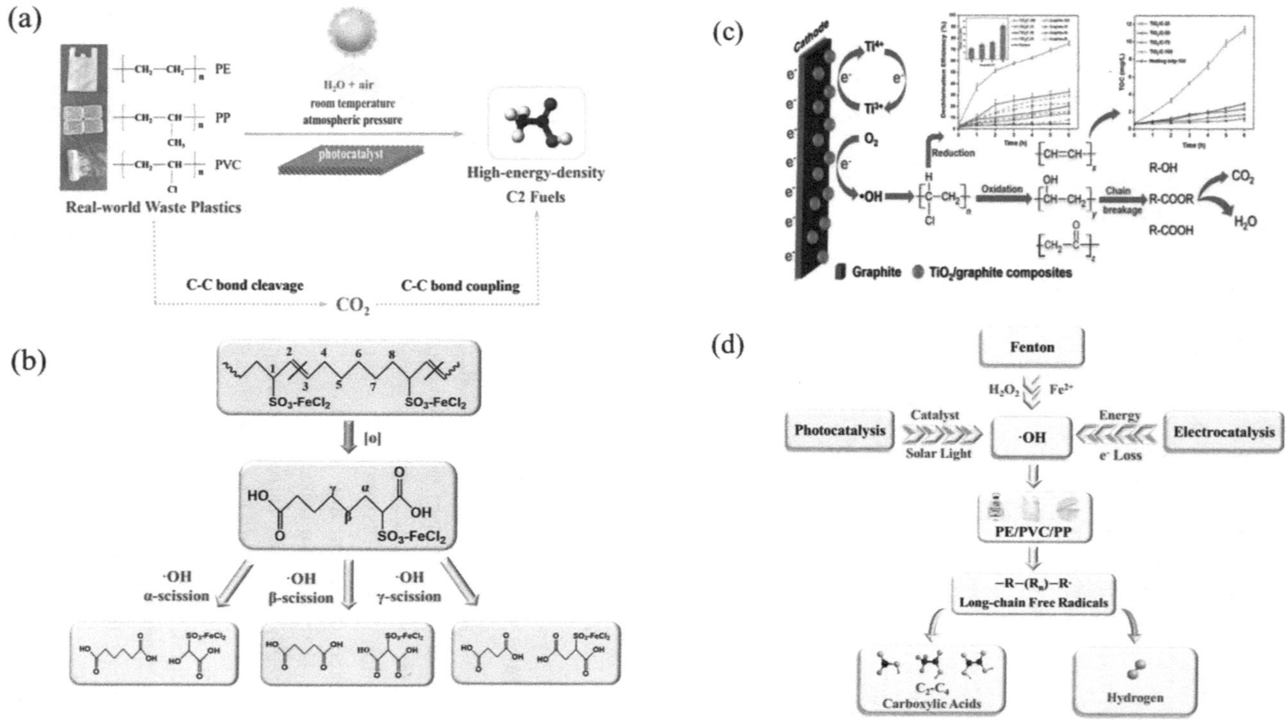

FIGURE 7.5 Mechanism of plastic conversion to fuels (a) Photocatalytic oxidation; (b) Fenton oxidation; (c) Electro-catalytic oxidation; (d) advanced oxidation. (Li et al. 2022.)

(Zeng et al. 2020; Kiran et al. 2017). The comprehension of the hydrogeology of the disposal site is crucial in order to achieve hydrogeological isolation. The selection and design of the site should prioritise the mitigation of water penetration and the prevention of toxins seeping into the surrounding environment. The consideration of long-term stability and durability is an integral aspect of the design and construction of disposal facilities. This ensures that the waste is effectively segregated and controlled for lengthy periods of time. The continuous monitoring and maintenance of disposal sites are vital in order to validate the effectiveness of barriers and guarantee the long-term containment of waste materials.

7.4 EMERGING PHYSICO-CHEMICAL TECHNIQUES FOR TECHNOLOGICAL WASTE VALORISATION

The preceding discussion elaborated on some industry-ready technologies and strategies to valorise and manage various technological waste streams. However, the push for a global circular economy and extensive research into green materials development and management has unearthed several cutting-edge physio-chemical strategies that primarily deal with the sustainable beneficiation of technological waste and extracting value from it. This section discusses a number of such relevant technologies and strategies for upcycling various technological waste streams.

7.4.1 Metal Extraction and Mining Waste

7.4.1.1 Electromagnetic Pulse Technology (EMPT)

The Electrochemical Metal Recovery Process (EMPT) is an innovative physico-chemical approach that aims to significantly transform the extraction of metals from mining waste (Marafi and Stanislaus 2008). The technology utilises the fundamental concepts of electromagnetic induction and selective heating in order to effectively extract precious metals. Electromagnetic induction is a process that utilises strong electromagnetic fields to generate eddy currents within waste materials that contain metal components (Wang et al. 2022). As the aforementioned currents traverse, they provide localised thermal energy, hence facilitating the segregation of metal particles from the waste matrix. Selective heating is a prominent benefit of electromagnetic pulse technology (EMPT) since it possesses the capability to heat and mobilise particular metals in a targeted manner. The attainment of selectivity is accomplished by the manipulation of the electromagnetic field's frequency and strength, enabling the effective segregation and retrieval of certain metals.

The utilisation of EMPT has a wide range of benefits in the fields of metal extraction and mining waste management. Resource recovery is a key aspect of EMPT, as it enables the extraction of metals from low-grade ores, tailings, and mining waste. This process not only helps to save important resources but also mitigates the necessity for resource-intensive mining operations. Environmental sustainability is also achieved by the utilisation of EMPT, which effectively reduces the negative environmental consequences often associated with conventional metal extraction methods, such as

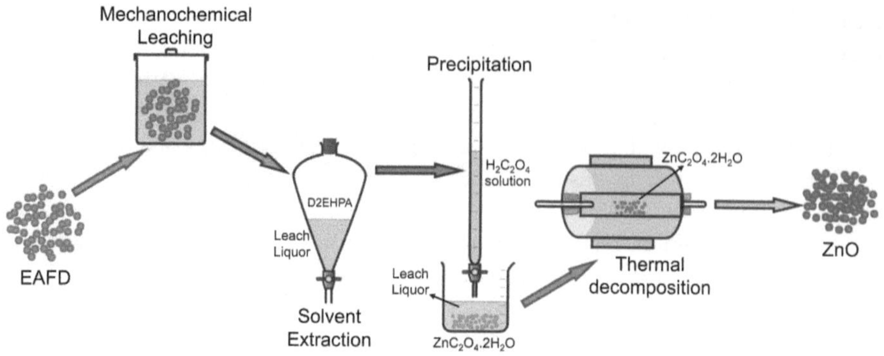

FIGURE 7.6 Schematic representation for recovery of Zn as ZnO from steelmaking waste materials by Mechanochemical Leaching, Solvent Extraction, Precipitation, and Thermal Decomposition Route. (Kalpakli et al. 2021.)

smelting and refining. Finally, the ability of EMPT to selectively extract individual metals enables the development of customised solutions that enhance recovery rates and reduce waste generation.

7.4.1.2 Mechanochemical Leaching

Mechanochemical leaching is an advanced physico-chemical methodology that improves the effectiveness of metal leaching from mining waste (Ou, Li, and Wang 2015; Tan and Li 2015). The approach is based on the fundamental concepts of mechanical activation and chemical leaching. Mechanical activation involves subjecting mining waste materials to mechanical forces, such as grinding, milling, or attrition, which serves the purpose of generating microcracks and augmenting the surface area that may be accessed by leaching agents (Odebiyi et al. 2022). Chemical leaching is then utilised, which involves the utilisation of specialised chemical agents that may be customised to selectively remove metals from waste materials by acting on their exposed surfaces. In **Figure 7.6**, a schematic depiction is shown of the recovery of zinc (Zn) in the form of zinc oxide (ZnO) from waste materials generated during the steelmaking process. This recovery process involves mechanochemical leaching, solvent extraction, precipitation, and thermal decomposition as key steps (Kalpakli et al. 2021).

Mechanochemical leaching presents several benefits in terms of metal extraction and the management of mining waste. The process of mechanical activation greatly improves the accessibility of metals, resulting in enhanced leaching efficiencies and better rates of metal recovery. Another advantage of mechanochemical leaching is its reduced energy consumption in comparison to conventional high-temperature processes like smelting. This characteristic contributes to the overall sustainability and cost-effectiveness of the process. Further, this methodology may be precisely customised to selectively target certain metals of interest, hence minimising waste and resource utilisation while optimising the recovery process.

7.4.2 Electronic Waste

7.4.2.1 Supercritical Fluid Extraction (SFE)

The new physico-chemical process, SFE, is employed to extract valuable elements from electronic waste, including metals, polymers, and several other components (**Figure 7.7**) (Li and Xu 2019). The efficacy of this approach is contingent upon the characteristics of supercritical fluids and the phenomenon of selective solvation. Supercritical fluids, shown as carbon dioxide (CO_2) in a supercritical state, demonstrate properties that encompass attributes of both gaseous and liquid phases (Li and Xu 2019; Preetam et al. 2023). These entities can permeate through solid structures akin to a gaseous state while concurrently dissolving materials in a manner reminiscent of a liquid. Such selective solvation refers to the process in which certain components of electronic waste are dissolved by supercritical CO_2, resulting in the separation of valuable elements from electronic waste that may be subsequently recovered (Chen et al. 2022).

SFE offers a multitude of benefits in the realm of electronic waste recycling and resource recovery. Selective extraction enables the targeted extraction of particular components, such as metals or polymers. This approach effectively minimises waste generation and enhances the efficiency of recycling procedures. The utilisation of supercritical CO_2 in extraction processes results in the generation of minimum residue, hence streamlining subsequent processing steps and mitigating environmental consequences (Li and Xu 2019).

7.4.2.2 Reactive Separation: Targeted Recovery of Valuable Metals

Reactive separation is a sophisticated physico-chemical methodology employed to selectively isolate precious metals from electronic waste. The process is dependent on chemical processes, such as leaching using ecologically benign acids, to dissolve and reclaim certain metals. Selective leaching is employed in reactive separation processes, where certain leaching agents are strategically employed to dissolve and remove precious metals from electronic waste while ensuring that non-target components remain unaffected. The subsequent stage after dissolution involves the precipitation and recovery of the targeted metals, which can be achieved through many methodologies, including electrodeposition or solvent extraction (Islam et al. 2021).

Reactive separation has a high level of selectivity, enabling the targeted recovery of certain metals while reducing waste and optimising resource efficiency. The utilisation of environmentally friendly leaching agents in the process mitigates the potential environmental consequences as compared to conventional methodologies that may include the usage of harmful chemicals (Cui and Zhang 2008). Resource conservation is another significant benefit of effective recovery and reuse of valuable metals in numerous sectors, as it helps to minimise the necessity for fresh resource extraction.

7.4.2.3 Ultrasound-Assisted Leaching: Enhancing Metal Recovery

Ultrasound-assisted leaching is a novel physico-chemical methodology employed to enhance the efficacy of metal extraction from electronic waste. The utilisation of ultrasonic waves are employed to augment the leaching process of precious metals

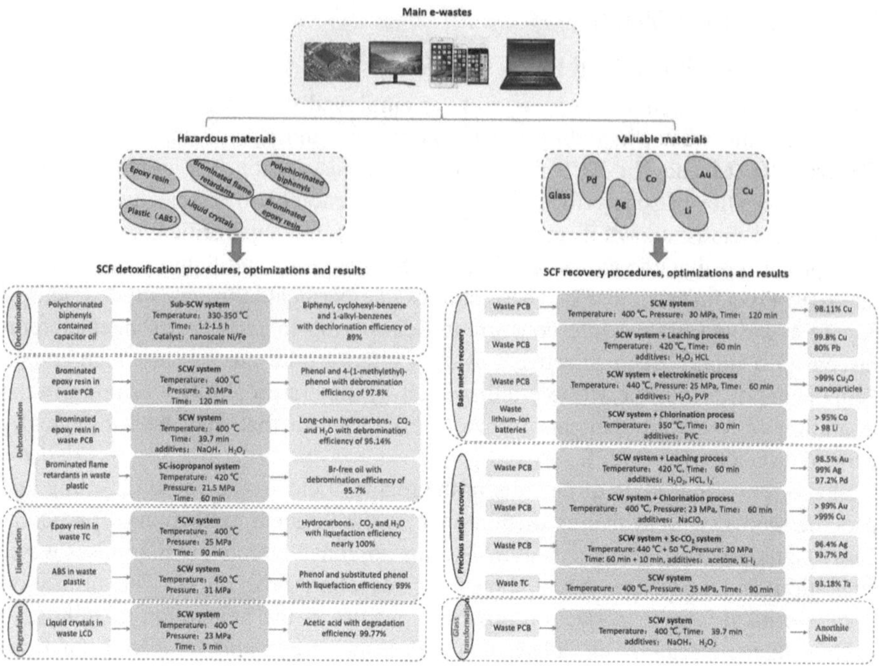

FIGURE 7.7 Detoxification processes of electronic waste by SFE technology to extract valuable metallic and non-metallic products. (Li and Xu 2019.)

from electronic components. The cavitation effect is observed when ultrasound waves induce the formation of tiny bubbles. The aforementioned bubbles undergo a forceful collapse, resulting in the production of concentrated heat, pressure, and microjets. This agitation facilitates increased interaction between the leaching solution and the surfaces of the metal (Bu et al. 2022). The application of ultrasound induces vigorous agitation, hence facilitating the extraction of metals from electronic waste components, leading to an increased efficiency in the leaching process (Li et al. 2017).

Ultrasound-assisted leaching has the ability to boost the efficiency of metal recovery processes. By facilitating improved interaction between leaching agents and metal surfaces, ultrasound effectively reduces both the time and costs associated with metal processing. The effectiveness of the procedure results in a decrease in the amount of leaching agent used, hence mitigating the environmental consequences associated with chemical consumption (Luque-García and Luque de Castro 2003). The extraction of valuable metals is conducted with high efficiency using ultrasound-assisted leaching, enabling their recycling and contributing to resource conservation and the reduction of electronic waste.

7.4.2.4 Ionic Liquids-Based Extraction: Green Solvents for Metal Recovery

The utilisation of ionic liquids as a means of extraction represents a physico-chemical method that is both ecologically sustainable and conducive to the retrieval of precious metals from electronic waste. Ionic liquids are a class of compounds characterised by

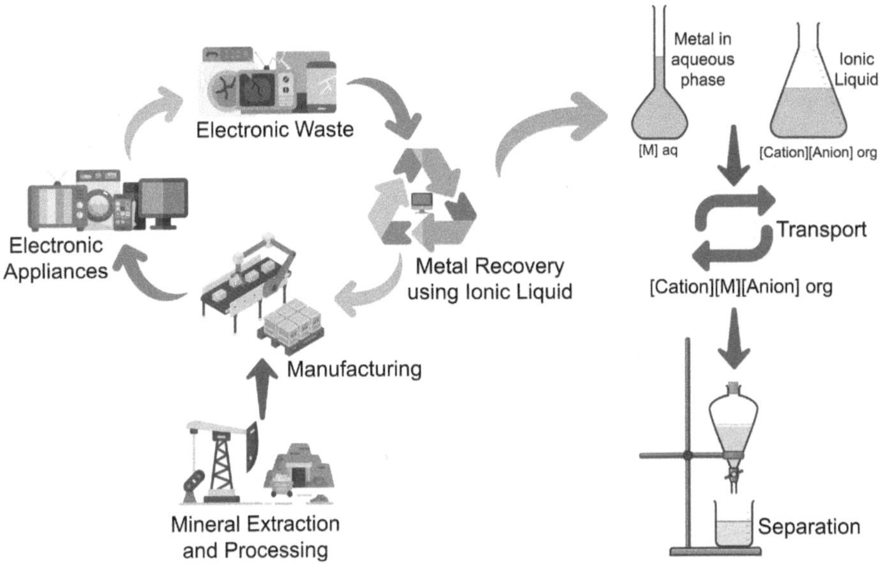

FIGURE 7.8 Schematic representation of ionic liquids in metal recovery from electronic waste and by liquid-liquid extraction. (AliAkbari et al. 2020.)

their salt-like nature and very low melting temperatures, often below 100°C. These substances possess the unique ability to be customised or modified in order to exhibit preferential solubility towards certain metal species (Barrueto et al. 2022). Ionic liquids can be tailored with distinct characteristics, enabling them to preferentially dissolve and retrieve desired metals while excluding other substances. Ionic liquids are also widely recognised as environmentally friendly solvents, primarily due to their low volatility and non-toxic properties. As a result, the utilisation of these solvents in metal recovery operations significantly mitigates the overall environmental effect. **Figure 7.8** illustrates the utilisation of ionic liquids for the purpose of extracting metals from electronic waste and secondary sources. The extraction process involves liquid-liquid extraction and electrodeposition techniques (AliAkbari et al. 2020).

Ionic liquid-based extraction provides several benefits for the recycling of electronic waste. Selective recovery enables the targeted extraction of precious metals, hence minimising waste generation and optimising the exploitation of resources. Ionic liquids exhibit a reduced environmental impact in comparison to conventional solvents, hence adhering to the principles of green chemistry (Yudaev and Chistyakov 2022). The efficient recovery and subsequent reuse of valuable metals in diverse applications contribute to the conservation of resources.

7.4.2.5 Biotechnological Recovery: Microbial Metal Extraction From E-Waste

The application of biotechnological methods in the management of electronic waste entails the utilisation of microorganisms or bioleaching techniques to extract precious metals from such waste in an environmentally sustainable manner (Islam et al. 2021).

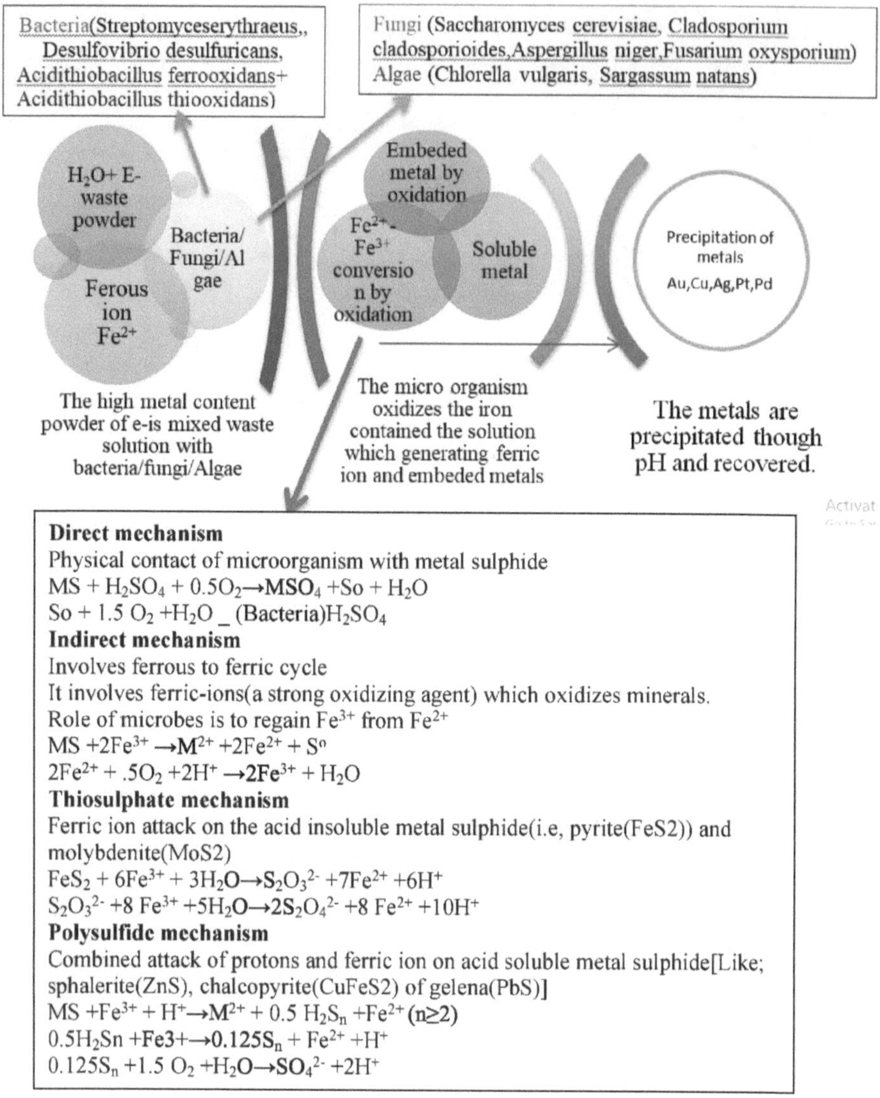

FIGURE 7.9 Biological mechanism for metal extraction from electronic waste. (Islam et al. 2020.)

Microbial activity may be used to facilitate the leaching and extraction of metals from electronic waste, with bacteria being a prominent example of such microorganisms. Bioleaching is an example that utilises microbial activity to break down and extract metals from electronic waste (Baniasadi et al. 2019). **Figure 7.9** shows the biosorption process, which encompasses precipitation, complexation, adsorption, and ion exchange, as a set of metabolically independent processes to extract several metals from electronic waste (Pant et al. 2012).

The utilisation of biotechnological methods in the recovery of electronic waste presents several benefits. The eco-friendly recovery strategy is predicated on harnessing natural microbial activities, mitigating the adverse environmental effects. Selective metal extraction uses certain microbes that can specifically target and extract precious metals (Arya and Kumar 2020).

7.4.3 Plastic Waste

7.4.3.1 Hydrothermal Liquefaction (HTL)

Hydrothermal Liquefaction (HTL) is a novel physico-chemical methodology employed for the purpose of plastic waste treatment. The process involves subjecting plastic waste to elevated temperatures and pressures while water is present in order to transform it into bio-crude oil and other useful by-products (Laredo, Reza, and Meneses Ruiz 2023). High-temperature liquefaction involves subjecting plastic waste to extreme temperatures (about 350–400°C) and pressures in the presence of water. In the given circumstances, plastics experience hydrothermal cracking, fragmenting their molecular structure into smaller hydrocarbon compounds. The main output of HTL is a kind of crude oil, which can undergo additional refining processes to be utilised as a viable energy source or as a useful feedstock for the production of chemicals (Dimitriadis and Bezergianni 2017).

HTL has several significant benefits in the context of plastic waste management. The process allows for extracting valuable hydrocarbons from plastic waste, offering a possible alternative for producing fuels and chemicals. The technology effectively decreases the volume of plastic waste, making a substantial contribution to waste reduction and the preservation of resources (Hu et al. 2021). The utilisation of bio-crude oil derived through HTL can serve as a viable energy source, therefore contributing to the advancement of sustainable practices.

7.4.3.2 Plasma Gasification: Transforming Plastic Waste Into Syngas

Plasma gasification is another advanced physico-chemical process that utilises a high-temperature plasma arc to disintegrate plastic waste, producing syngas, a combination of hydrogen and carbon monoxide. The aforementioned procedure entails the utilisation of temperatures that surpass 3,000°C. Plasma generation initiates a plasma discharge by applying high electrical energy to a gas, often air or pure oxygen. This phenomenon gives rise to an extremely reactive setting characterised by temperatures that exceed those observed on the surface of the sun. Gasification involves the conversion of plastic waste into syngas through the plasma state when the organic constituents undergo decomposition. Simultaneously, the inorganic components transform into a glassy material known as slag (Munir et al. 2019).

Energy recovery is a significant benefit of the process, as the syngas generated may be effectively utilised to create energy, such as producing electricity or heat. This promotes the conversion of waste into a valuable energy source. The use of this approach also results in a significant reduction in environmental impact. This is achieved by utilising high temperatures, which facilitate the full combustion of waste materials. Additionally, the syngas produced may be effectively caught and

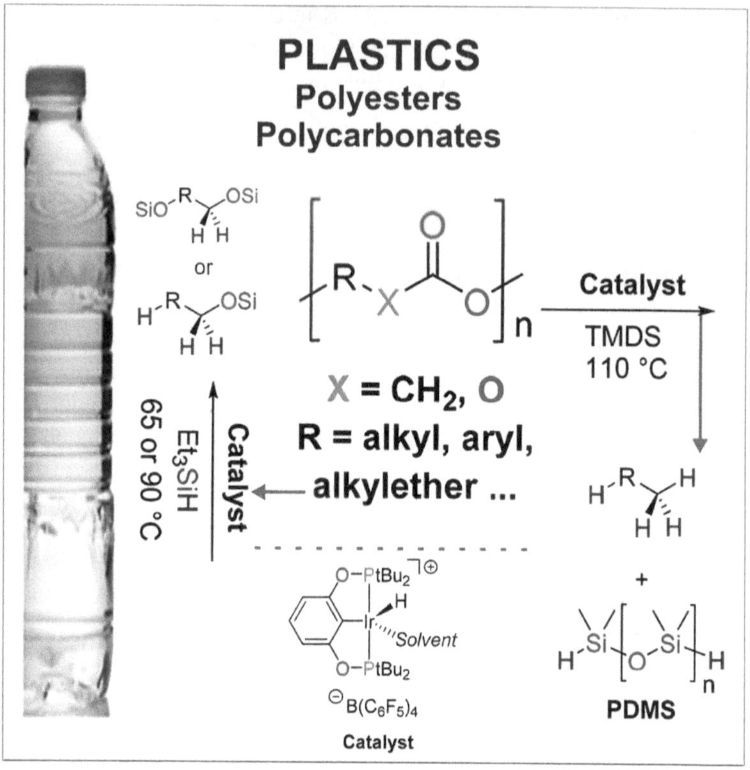

FIGURE 7.10 Depolymerisation of waste plastics to monomers and chemicals. (Monsigny, Berthet, and Cantat 2018.)

processed, further enhancing the efficiency of the process and minimising emissions (Sanlisoy and Carpinlioglu 2017).

7.4.3.3 Solvent-Based Depolymerisation

Solvent-based depolymerisation is a physico-chemical methodology that uses specialised solvents to facilitate the disintegration of polymers into their constituent monomers (Valerio, Muthuraj, and Codou 2020). Subsequently, these monomers can be utilised to synthesise novel plastics or other materials of significant worth (**Figure 7.10**) (Monsigny, Berthet, and Cantat 2018). The selection of solvents is critical as they can selectively dissolve and depolymerise the plastic substance. Monomer recovery involves separating and subsequently using the depolymerised monomers for diverse purposes.

Solvent-based depolymerisation presents several advantages in the context of plastic waste management. Closed-loop recycling involves using the recovered monomers to manufacture new polymers, hence facilitating the promotion of closed-loop recycling. The utilisation of monomers in material upcycling processes facilitates the production of more valuable goods, hence diminishing the dependence on primary resources. Through this method, waste diversion effectively redirects plastics away from landfills or incineration processes, significantly contributing to the overall decrease in plastic waste (Sherwood 2020).

7.4.3.4 Enzymatic Degradation

Enzymatic degradation is a physico-chemical process that employs enzymes as catalysts to facilitate the decomposition of polymers into less complex molecules. This concept presents a more ecologically sustainable strategy for managing plastic waste (Roohi et al. 2017). Enzyme specificity refers to the ability of different enzymes to selectively target various kinds of polymers for breakdown. The enzymatic degradation process is considered eco-friendly due to its little environmental effect and its ability to proceed at lower temperatures (Mohanan et al. 2020).

Enzymatic degradation presents several advantages in the realm of plastic waste management. The method exhibits a high degree of environmental friendliness since it abstains from the use of harsh chemicals or elevated temperatures. Selective degradation, another advantage, which involves the deliberate selection of enzymes that can specifically target certain forms of plastic, enhancing the effectiveness of recycling processes. The biodegradable nature of the product results in the production of end products that can undergo decomposition, hence mitigating the potential for prolonged environmental damage (Kaushal, Khatri, and Arya 2021).

7.4.3.5 Microwave-Assisted Pyrolysis

Microwave-assisted pyrolysis uses microwave energy to facilitate the pyrolysis process, which involves decomposing plastic waste into valuable chemicals and fuels (Arshad et al. 2017). Microwave heating has been found to be an effective method for facilitating the degradation of polymers in plastic waste and facilitates the fast conversion of plastic waste. **Figure 7.11** shows the process of catalytic pyrolysis of plastic waste within a continuous microwave-assisted pyrolysis system with the objective of producing fuel (Zhou et al. 2021).

The rapid processing technology of microwave-assisted pyrolysis demonstrates a substantial decrease in processing time when compared to conventional pyrolysis procedures. Energy recovery is a significant aspect of the process since it yields valuable chemicals and fuels, positively contributing to sustainability. The process of converting plastic waste into valuable chemicals has the potential to decrease the reliance on primary resources (Arshad et al. 2017).

7.4.4 Industrial Chemical Waste

7.4.4.1 Supercritical Water Oxidation (SCWO)

Supercritical Water Oxidation (SCWO) is a sophisticated physico-chemical methodology utilised for the treatment of hazardous chemical waste. The system functions under elevated temperature and pressure, employing supercritical water as a medium for the oxidation and decomposition of organic and hazardous substances (Schmieder and Abeln 1999). Distinct features, such as a notable solvating capacity and accelerated chemical reaction speeds, characterise water's characteristics in the

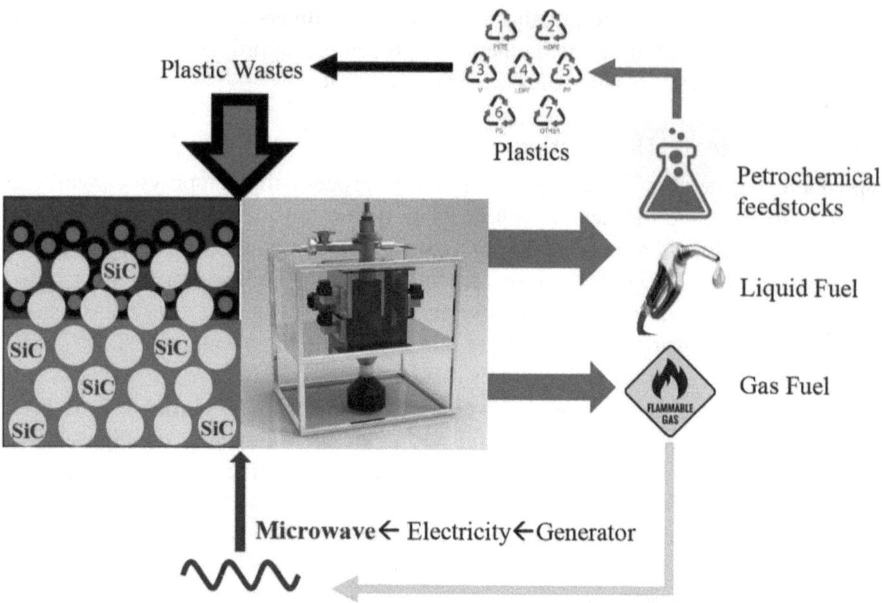

FIGURE 7.11 Catalytic pyrolysis of plastic waste using continuous microwave assisted pyrolysis. (Zhou et al. 2021.)

supercritical state. This chemical exhibits efficient oxidation of organic compounds. SCWO uses the potent oxidising capabilities of supercritical water to decompose complex organic molecules into harmless by-products, predominantly carbon dioxide and water (Marrone 2013).

SCWO exhibits a high degree of efficacy in decomposing intricate and perilous organic compounds, guaranteeing the thorough eradication of pollutants. The utilisation of this approach results in a reduction in the production of secondary waste materials, hence streamlining the waste treatment procedure. SCWO also produces thermal energy as a by-product of the oxidation reaction, which may be harnessed and utilised for many applications, hence facilitating the advancement of sustainable practices (Tang et al. 2021).

7.4.4.2 Advanced Oxidation Processes (AOPs): Targeted Treatment of Hazardous Chemicals

Advanced Oxidation Processes (AOPs) refer to a collection of physico-chemical methodologies that employ potent oxidants such as ozone, hydrogen peroxide, or UV radiation to disintegrate and decontaminate dangerous chemical waste (Deng and Zhao 2015). The aforementioned mechanisms are dependent on the production of very reactive hydroxyl radicals in order to launch oxidation reactions. Hydroxyl radical generation is facilitated by AOPs, which employ several mechanisms, including photocatalysis, ozonation, and Fenton-like reactions. These processes effectively produce hydroxyl radicals, which are recognised as highly reactive oxidative species.

The oxidation mechanisms involve the assault of hydroxyl radicals on both organic and inorganic pollutants, resulting in their oxidation and subsequent degradation into simpler molecules that pose reduced hazards (Comninellis et al. 2008).

AOPs have several benefits in the context of hazardous chemical waste management. Selective destruction refers to the ability of AOPs to be customised in order to specifically target and eliminate particular impurities. This capability ensures the effective and efficient elimination of dangerous compounds. The utilisation of hydroxyl radicals' oxidative capabilities enables the reduction of chemical doses, hence mitigating the environmental consequences associated with chemical waste treatment. The versatility of AOPs allows for their application in a diverse array of chemical pollutants, rendering them highly adaptable instruments for the treatment of hazardous waste (M'Arimi et al. 2020).

7.4.4.3 Photochemical Degradation

Photochemical degradation utilises ultraviolet (UV) or visible light, along with photocatalysts, to facilitate the breakdown and detoxification of toxic organic compounds found in chemical waste (Kumar et al. 2020). Photocatalysts, often semiconductor materials such as titanium dioxide (TiO_2), can absorb photons emitted by light and produce electron-hole pairs. These very reactive entities engage in redox reactions with pollutants. Photocatalysis serves as a catalyst for oxidative processes, facilitating the degradation of dangerous organic molecules into by-products that are either less toxic or non-toxic in nature (Rueda-Marquez et al. 2020).

Photochemical degradation is characterised by its ecologically benign nature since it functions in mild settings and does not necessitate the utilisation of harmful chemicals. Selective degradation refers to the ability of photocatalysis to specifically target and break down particular organic pollutants, hence improving the overall efficiency of the treatment process. The utilisation of solar energy in photochemical degradation processes can effectively reduce energy consumption and contribute to the promotion of sustainable practices (Xie et al. 2021).

7.4.4.4 Wet Air Oxidation (WAO)

Wet Air Oxidation (WAO) is a physico-chemical process employed to treat chemical waste that contains organic molecules posing potential hazards. The system functions in an environment characterised by increased temperature and pressure while immersed in an aqueous medium (Kurniawan, Lo, and Chan 2006). WAO entails the oxidation of organic compounds in the presence of water and oxygen. This reaction occurs at temperatures ranging from 150°C to 320°C and pressures ranging from 20 to 220 bar. The process of oxidative decomposition occurs when organic substances undergo oxidation and decomposition under circumstances of high temperature and pressure. This results in the formation of simpler by-products that are less dangerous, namely carbon dioxide and water (Villegas et al. 2016).

WAO proficiently disassembles intricate organic molecules, guaranteeing their thorough degradation. The approach employed in WAO effectively mitigates the environmental impact by minimising the formation of secondary waste products, hence simplifying the subsequent waste treatment procedures. The process of WAO results in the production of heat energy, which may be harnessed and utilised for many applications, contributing to the advancement of sustainable practices (Capodaglio, Hlavínek, and Raboni 2015).

7.4.4.5 Fenton Reaction: Oxidative Treatment for Hazardous Chemicals

The Fenton reaction is a physico-chemical process employed to remedy hazardous chemical waste (Bautista et al. 2008). The process entails the utilisation of hydrogen peroxide (H_2O_2) and iron catalysts to produce exceptionally reactive hydroxyl radicals in order to facilitate the oxidative destruction of organic chemical pollutants. Hydroxyl radical generation occurs through the Fenton reaction, wherein iron catalysts, commonly in the form of ferrous ions (Fe^{2+}), undergo redox interactions with hydrogen peroxide, producing hydroxyl radicals. Oxidative decomposition involves the utilisation of hydroxyl radicals, which possess strong oxidising properties. These radicals serve as initiators for oxidative decomposition events, leading to the breakdown of organic chemical pollutants into molecules that are simpler in structure and provide reduced levels of danger (Babuponnusami and Muthukumar 2014).

The Fenton reaction presents several advantages in the context of hazardous chemical waste management. The hydroxyl radicals produced during the Fenton reaction exhibit a high level of reactivity, making them exceptionally powerful oxidative agents that effectively facilitate the destruction of contaminants. The approach exhibits versatility in its applicability to diverse organic pollutants, hence rendering it suitable for treating hazardous waste. The use of hydroxyl radicals' oxidative capabilities enables the reduction of chemical doses, hence mitigating the environmental consequences associated with chemical waste treatment (Jiang and Zhang 2007).

7.4.5 Nuclear Waste

7.4.5.1 Partitioning and Transmutation (P&T)

The development of nuclear power offers a considerable supply of environmentally friendly energy. Nevertheless, it also generates radioactive waste with long-lasting properties, which presents notable difficulties in terms of disposal and ensuring environmental safety. The protracted radiotoxicity and possible risks associated with long-lived radioactive isotopes, such as plutonium-239 and certain actinides, pose significant challenges. Partitioning and Transmutation (P&T) has emerged as a technologically sophisticated and promising approach to address these difficulties. The concepts behind the operation of P&T involve the selective separation and transmutation of radioactive isotopes, intending to efficiently manage long-lived isotopes. The initial stage of the process entails the discerning isolation of long-lived radioactive isotopes from the intricate nuclear waste mixture. Different separation techniques are utilised in various applications, such as solvent extraction, ion exchange chromatography, and adsorption procedures. These techniques utilise the distinctive chemical and physical characteristics of particular isotopes in order to separate them from the surrounding waste material. Transmutation is a subsequent step following the separation of radioactive isotopes when these isolated isotopes are exposed to neutron bombardment within specialised reactors. The objective of transmutation is to transform isotopes that are both long-lived and dangerous into species that are either shorter-lived or less radioactive (Salvatores and Palmiotti 2011). This method can alter the nuclear characteristics of isotopes, resulting in a gradual reduction of their hazardous nature.

P&T has several significant benefits in the context of nuclear waste management. The reduction of radiotoxicity is achieved by the process, which selectively targets and transmutes long-lived radioactive isotopes present in nuclear waste. The process of transmutation results in a decrease in both the volume and radiotoxicity of nuclear waste, hence enhancing its storage and disposal feasibility. This approach also has the potential to mitigate the necessity of constructing deep geological repositories that require extensive confinement periods (Bourg et al. 2020). The use of P&T techniques diminishes the radiotoxicity and lifespan of nuclear waste, hence decreasing the potential environmental and health hazards linked to nuclear waste dumps. It significantly improves the safety of storing waste over extended periods of time.

Although the field of P&T exhibits considerable potential, it is not without obstacles. Implementing separating and transmutation technologies in the field of P&T necessitates utilising complex technical processes. These processes include substantial investments in research, development, and infrastructure. The development of specialised nuclear reactors for the purpose of efficient transmutation poses a significant technological and engineering challenge. Safety and regulatory considerations are also crucial in P&T processes and facilities due to their handling of highly radioactive materials (Kurniawan et al. 2022).

7.4.5.2 Electrochemical Separation: Targeted Removal of Radioactive Isotopes

Electrochemical separation is a sophisticated physico-chemical methodology employed to selectively isolate distinct radioactive isotopes from nuclear waste streams. This phenomenon is predicated upon the fundamental principles of electrochemistry and the utilisation of ion-selective membranes. Ion-selective membranes are utilised in electrochemical separation systems to enable the selective transit of some ions while impeding the passage of others. Selective isotope removal may be achieved by the utilisation of electrochemical separation, which involves the manipulation of ion-selective membranes and the application of voltage (Kurniawan et al. 2022). This technique enables the targeted extraction of radioactive isotopes from nuclear waste solutions.

The method of selective isotope removal enables the deliberate elimination of particular radioactive isotopes, decreasing the waste material's radiotoxicity. Efficient resource recovery entails the retrieval of valuable isotopes from nuclear waste, intending to potentially repurpose them for applications in the fields of medicine, industry, or scientific research. Electrochemical separation has the advantage of reducing the bulk of nuclear waste, facilitating long-term storage and disposal (Trojanowicz, Kołacińska, and Grate 2018).

7.4.5.3 Advanced Sorbents: Efficient Capture of Radioactive Materials

Advanced sorbents refer to specialised materials that have been specifically engineered to effectively collect and adsorb radioactive substances present in solutions containing nuclear waste. These materials frequently exhibit a high degree of selectivity towards particular radionuclides (Li et al. 2018; Hao et al. 2023). This enables the targeted removal of these radionuclides from solutions containing nuclear waste. The immobilisation process involves confining radioactive elements within the sorbent material, effectively limiting their potential release into the surrounding environment (Riley et al. 2016).

Advanced sorbents can selectively collect and remove certain radionuclides, resulting in efficient removal processes. The immobilisation of radioactive elements within the sorbent significantly mitigates the potential for environmental contamination, hence reducing the overall environmental impact. The utilisation of improved sorbents has the potential to enhance the process of resource recovery by enabling the retrieval and subsequent reuse or recycling of important radioactive materials (Manos and Kanatzidis 2016).

7.4.5.4 Hybrid Sorption-Ion Exchange

The hybrid sorption-ion exchange method is a physico-chemical technology that integrates both sorption and ion exchange processes in order to efficiently absorb and eliminate radioactive substances from solutions containing nuclear waste. The sorption process involves the selective capture of radioactive elements by a solid sorbent substance that has a strong affinity for certain radionuclides (Phillip et al. 2023). The sorbent material is next exposed to an ion exchange procedure, in which the radioactive ions caught are replaced with less harmful ions, therefore facilitating their removal.

The selective removal approach enables the targeted elimination of certain radioactive substances, therefore decreasing the radiotoxicity of the waste. Hybrid systems have demonstrated the ability to effectively capture radioactive materials, resulting in high rates of removal. The technique exhibits versatility as it may be modified to effectively target a diverse array of radionuclides, hence rendering it suitable for nuclear waste treatment over a broad spectrum (Rahman, Ibrahium, and Hung 2011).

7.4.5.5 Reductive Decontamination

The reductive decontamination process involves using physico-chemical methods to decrease the oxidation state of radioactive substances, hence diminishing their mobility and potential hazards. The process utilises reducing agents in order to chemically convert radionuclides. Reduction reactions encompass the process of reductive decontamination, wherein chemical reactions occur involving the transfer of electrons from reducing agents to radionuclides, therefore inducing a change in their oxidation state (Yang, Park, and Lee 2018). The immobilisation process involves minimising the mobility and environmental effects of radionuclides by reducing their concentration.

The utilisation of reductive decontamination presents several benefits in the realm of nuclear waste management. Through the process of lowering the oxidation state of radionuclides, this method effectively decreases their level of danger and diminishes the probability of their dispersion within the surrounding environment. The enhanced long-term stability of immobilised radionuclides plays a significant role in ensuring the safety of nuclear waste sites (Pabby et al. 2022). Resource conservation can be achieved by the implementation of reductive decontamination techniques, which have the potential to aid in the retrieval of valuable components from nuclear waste.

7.5 BIOLEACHING AND BIOREMEDIATION

Among the several discussed physico-chemical technological waste management techniques, biological-based resource valorisation techniques stand out because of their environmentally beneficial and sustainable characteristics. These approaches utilise natural processes and microorganisms to facilitate the decomposition of waste materials, therefore circumventing the need for the use of potentially dangerous chemicals or energy-intensive operations. They demonstrate exceptional proficiency in the field of resource recovery, namely in the extraction of precious metals from ores through a process known as bioleaching, as well as in the removal of pollutants through bioremediation. These capabilities hold significant potential for facilitating material recycling endeavours (Cui and Zhang 2008). They have been found to produce a reduced quantity of secondary wastes, exhibit adaptability to various waste compositions, and function for prolonged durations, hence mitigating long-term environmental hazards. Furthermore, these solutions have the potential to be economically efficient and in line with sustainability objectives, thus reducing environmental impacts and supporting circular economy aims (Olson, Brierley, and Brierley 2003). Therefore, it is pertinent to discuss them in further detail.

Bioleaching is a biologically mediated method that utilises the metabolic capabilities of microorganisms to dissolve metals from ores and diverse waste streams (Arya and Kumar 2020). This technique has several benefits, including a reduction in the environmental effect compared to traditional mining methods, a decrease in waste volume, and frequently an enhancement in metal recovery rates. The fundamental tenets of bioleaching encompass the involvement of microorganisms, the process of metal solubilisation, and the subsequent retrieval of metals.

- **Microbial Activity:** The process of bioleaching depends on the presence of certain microorganisms, often extremophiles, that can flourish in very challenging settings characterised by levels of acidity or salinity. The microorganisms possess a distinctive capability to undergo oxidation of metal sulfides, transforming them into metal ions soluble in solution (Baniasadi et al. 2019).
- **Metal Solubilisation:** Microbial oxidation produces metabolic by-products, such as sulfuric acid and organic acids, which contribute to the solubilisation of metals (Rasoulnia et al. 2021). The aforementioned acids undergo a chemical reaction with metal sulfides, resulting in the liberation of metal ions that subsequently dissolve in the surrounding solution.
- **Metal Recovery:** Following the solubilisation of metal ions, their retrieval can be accomplished using many methodologies, including precipitation, ion exchange, or electroplating, contingent upon the particular metal and its intended application.

There are several advantages associated with the process of bioleaching.

- **Environmental Sustainability**: Bioleaching is a process that promotes environmental sustainability by minimising the reliance on harsh chemicals and energy-intensive techniques, hence sustainably facilitating the management of resources.
- **Selective Metal Recovery**: Bioleaching can specifically target and recover certain metals from waste streams, diminishing the total creation of waste and fostering the conservation of resources.

The investigation of the interaction between microorganisms and waste materials has been conducted through bioleaching. Additionally, there is another variant known as spent medium bioleaching, where there is no direct contact between the microbes and the waste, as depicted in **Figure 7.12** (Arya and Kumar 2020). However, the bioleaching process may exhibit a reduced rate of extraction compared to chemical leaching techniques, thereby impacting the overall efficacy and economic viability of metal retrieval.

Bioleaching has versatility in its potential application in beneficiating different technological waste streams. The use of bioleaching enables the extraction of precious metals, including gold, silver, and palladium, from electronic waste (Udayakumar, Razak, and Ismail 2022). One prominent example of the utilisation of bioleaching is in the restoration of the Iron King Mine and Humboldt Smelter Superfund site in

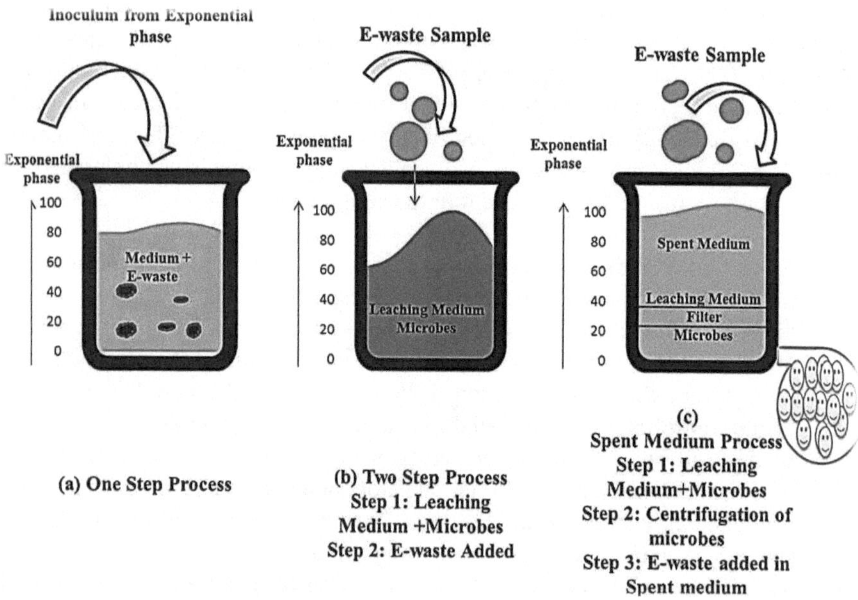

FIGURE 7.12 Variations and different steps of bioleaching of electronic waste. (Arya and Kumar 2020.)

Arizona. Bioleaching procedures are also employed to mitigate the concentrations of heavy metals in both soil and water samples. The utilisation of indigenous extremophilic microorganisms is employed to augment the process of metal removal and mitigate the adverse effects of environmental pollution (Xu et al. 2019; Yaashikaa et al. 2022). The utilisation of bioleaching can also contribute to the remediation of chemical waste through the degradation of perilous organic compounds. The use of bioleaching methods has also been utilised to detoxify soil that has been polluted with organophosphate pesticides. The introduction of microorganisms possessing the capacity to decompose these pesticides has been used as a means to augment the breakdown of pollutants and facilitate the restoration of locations that have been polluted. One of the benefits of this approach is that it offers an environmentally viable means of detoxifying chemical waste and facilitating ecosystem restoration. The utilisation of bioleaching has the potential to facilitate the targeted extraction of certain radionuclides from nuclear waste (Akash et al. 2022). Ongoing research is now exploring the potential application of bioleaching in the selective removal of uranium and other radionuclides from nuclear waste solutions. This technique aims to mitigate the radiotoxicity of nuclear waste products by harnessing microbial interactions with these pollutants.

Like bioleaching, **bioremediation** is an environmentally friendly technique that employs microbes, plants, or enzymes to facilitate the degradation or detoxification of harmful compounds in polluted ecosystems (Ramírez-García, Gohil, and Singh 2019). The fundamental scientific concepts behind bioremediation encompass microbial metabolism, methods of contaminant breakdown, and the restoration of the environment. Microbial metabolism involves using contaminants by microorganisms as energy sources or substrates, resulting in their subsequent modification or decay. Microorganisms can enzymatically break down dangerous substances, producing by-products that are either less toxic or non-toxic through metabolic processes (Medfu Tarekegn, Zewdu Salilih, and Ishetu 2020). Environmental restoration is the primary objective of bioremediation, which aims to return damaged habitats to their original or secure condition by reducing or eliminating pollutant concentrations.

The benefits of bioremediation are multifaceted. Bioremediation is an ecologically conscious and sustainable method for remediating polluted ecosystems, hence reducing reliance on chemical-based remediation techniques. Selective pollutant removal, a tenet of bioremediation, refers to the ability to specifically target and degrade particular pollutants without harming the surrounding environment. Bioremediation can provide long-term control and mitigation of pollution, facilitating ecosystem recovery and reducing contamination risks. Nonetheless, there are several drawbacks associated with the use of bioremediation techniques. The efficacy of bioremediation is contingent upon a range of factors particular to the site, including the nature of the pollutants and the availability of appropriate microbial species. Consequently, the effectiveness of this process might vary (Iwamoto and Nasu 2001). The duration of treatment in bioremediation procedures may be longer than alternative approaches, particularly when addressing complicated pollutants or expansive locations (Singh et al. 2017).

Bioremediation is a widely applicable technique that is utilised in many technological waste streams. Implementing bioremediation techniques has the potential to alleviate the adverse environmental effects associated with mining waste by detoxifying soils and rivers polluted with metals. The Iron King Mine and Humboldt Smelter Superfund site in Arizona underwent a case study on the application of bioremediation to rehabilitate the area and mitigate heavy metal contamination in the surrounding ecosystem (Menka, Root, and Chorover 2014). Bioremediation also has the potential to selectively address the presence of harmful organic chemicals found in electronic waste, such as flame retardants. Several researchers have investigated the utilisation of microbial strains that can degrade brominated flame retardants often present in electronic waste constituents. Bioremediation facilitates the microbial decomposition of dangerous compounds present in chemical waste, which is also well documented, leading to their detoxification.

7.6 PLASMA REMEDIATION

Like bioleaching and bioremediation, Plasma remediation is an innovative physicochemical technology that employs high-energy plasma (instead of microbes) to treat and convert diverse forms of technological waste (Gomez et al. 2009). Plasma remediation is founded upon developing plasma chemical processes and transforming waste materials. **Figure 7.13** illustrates the typical workflow for the utilisation of plasma remediation technology in the treatment of electronic waste to generate various functional materials (Sanito, You, and Wang 2021). The generation of plasma involves the utilisation of either elevated temperatures or the application of an electric field to a gas medium. As a consequence, the development of a state of matter characterised by high energy and ionisation occurs. Chemical reactions occur in the plasma state when gas molecules and atoms exhibit heightened reactivity, enabling the breakdown and transformation of waste components (Aggelopoulos 2022). Plasma remediation can facilitate the decomposition of intricate organic compounds, the ionisation of metals, and the breakdown of hazardous contaminants. Consequently, this process enables the conversion of waste materials into less complicated and less detrimental forms. Plasma remediation has demonstrated remarkable efficacy in the process of breaking down and converting diverse waste materials, therefore mitigating their adverse environmental effects. Selective pollutant removal refers to the process of specifically targeting and eliminating certain pollutants, such as heavy metals, organic compounds, and radioactive elements. Plasma remediation procedures are also characterised by their rapidity, which allows for efficient waste treatment and reduces the potential for environmental exposure. However, producing high-energy plasma necessitates a significant amount of energy input, amplifying operating expenses and environmental impact. The design, construction, and operation of plasma remediation systems may be intricate and expensive (Levchenko et al. 2022).

Despite these, plasma remediation finds uses in valorising various technological waste streams. The utilisation of plasma remediation has the potential to mitigate the environmental consequences associated with mining waste through the process of detoxifying soils polluted with heavy metals and addressing the issue of acid mine

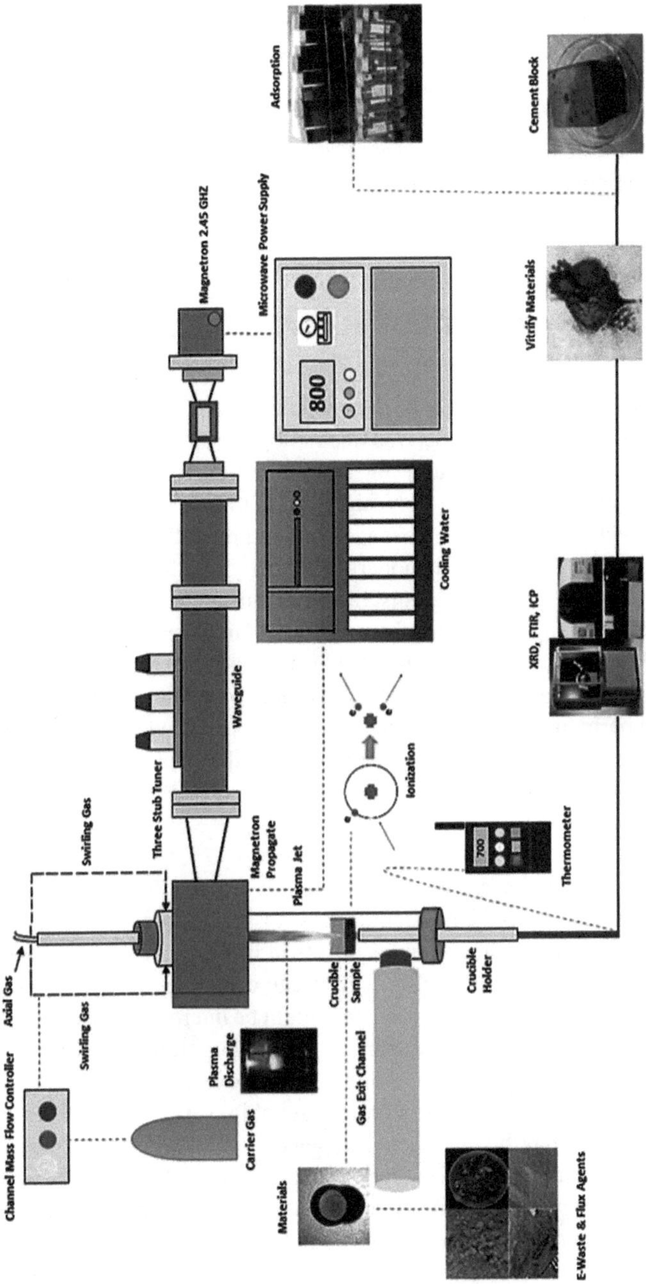

FIGURE 7.13 Workflow depicting the application of plasma technology for treating e-waste. (Sanito, You, and Wang 2021.)

drainage. This treatment method provides prompt remediation, efficient immobilisation of heavy metals, and mitigates the discharge of contaminants. The application of plasma technology has been employed to remediate soil polluted with heavy metals at abandoned mining sites in the United States. The principle of plasma remediation on electronic waste involves the extraction of precious metals from electronic waste and decomposing harmful organic molecules. The application of plasma technology in recovering metals such as copper and aluminium from electronic waste is of much scientific interest. Plasma remediation also possesses the capability to disintegrate intricate organic compounds that are commonly present in chemical waste streams. This allows plasma remediation to treat chemical waste that consists of volatile organic compounds (VOCs) and persistent organic pollutants (POPs) (Zhang et al. 2017).

7.7 NANOTECHNOLOGY FOR SUSTAINABLE REMEDIATION OF TECHNOLOGICAL WASTE

The latest and more exciting prospect in technological waste management is the advent of nanotechnology and derived technologies as significant and influential drivers, presenting novel and enduring resolutions for mitigating and valorising technological waste. Nanotechnology signifies a transformative paradigm within the domain of technological waste management and remediation (Dutta et al. 2018). The distinctive characteristics of nanoparticles, including their heightened reactivity and selectivity, provide viable and sustainable solutions for a wide range of waste streams, including electronic waste, chemical waste, nuclear waste, and plastic waste (Serag 2022; Lagaron and Lopez-Rubio 2011). The unquestionable potential of nanotechnology lies in its capacity to convert waste into valuable resources and mitigate environmental harm, notwithstanding the problems associated with scalability, environmental implications, and regulatory frameworks that need to be addressed. The rationale for employing nanotechnology in the context of technological waste management is multifaceted.

- The reactivity and surface area of materials are significantly enhanced in the nanoscale, owing to the substantial ratio of surface area to volume. This distinctive attribute of nanoparticles results in increased reactivity, hence enhancing their efficacy in engaging with waste constituents. Within the realm of waste management, this phenomenon might be interpreted as a more expedited process of eliminating or transforming pollutants (Jain et al. 2021).
- Engineered nanoparticles may be customised to exhibit unique affinity towards certain contaminants. By including suitable surface groups or coatings, nanoparticles may be engineered to selectively trap certain pollutants while disregarding harmless substances (Khan et al. 2021). The aforementioned selectivity enhances the effectiveness of waste treatment procedures.
- The phenomenon of catalysis and redox activity may be observed in some types of nanoparticles, which possess inherent features that enable them to facilitate chemical reactions and participate in oxidation-reduction processes. Catalysts can expedite chemical processes, such as degrading organic

contaminants or reducing hazardous metals. The catalytic properties of these substances facilitate the conversion of noxious waste constituents into less deleterious ones.

- The use of nanotechnology also facilitates the retrieval of valuable materials from various waste streams. For example, it has the potential to enhance the effective retrieval of valuable metals from electronic waste or the breakdown of polymers into their individual monomers, thus supporting the concept of a circular economy and mitigating the depletion of resources (Kumari and Samadder 2022).

The adaptability of nanotechnology is exemplified by its wide range of applications in the remediation of various technological waste streams. The utilisation of nanotechnology is crucial in the process of recycling and extracting precious metals from electronic waste (Kumari and Samadder 2022). Nanoscale materials, such as carbon nanotubes and nanoparticles, are employed in order to optimise the efficacy of processes such as leaching and electrochemical recovery. Several advantages are associated with this approach, including enhanced rates of metal recovery, less environmental effect, and reduced formation of waste. Nanoparticles are also utilised to degrade and detoxify harmful substances present in chemical waste. Photocatalytic nanoparticles, such as those mentioned, enable the degradation of organic contaminants by light-induced processes. There are several advantages associated with this approach. Firstly, it is environmentally benign, as it promotes sustainable practices and minimises harm to the ecosystem. Secondly, it allows for selective removal of pollutants, enabling targeted treatment of specific contaminants. Lastly, it reduces dependence on chemical treatments, which may negatively impact human health and the environment. Besides these applications, nanotechnology provides viable approaches for containing and isolating radionuclides within nuclear waste. Engineered nanoparticles can establish stable matrices, reducing the mobility and radiotoxicity of radioactive substances (Salwadi and Kurapati 2022). There are several advantages associated with this approach. Firstly, it enhances radiological safety, reducing the potential risks and hazards associated with handling and storing radioactive materials. Additionally, it reduces waste volume, which is beneficial in terms of storage and disposal. Lastly, it offers enhanced long-term storage alternatives, ensuring the safe and secure containment of radioactive materials over extended periods. Nanotechnology also plays a significant role in facilitating the degradation and recycling of plastic waste. The utilisation of nanocatalysts and enzymes can accelerate the depolymerisation process of plastics into their constituent monomers. There are several advantages associated with this initiative. Firstly, it serves to promote the recycling of plastic materials, contributing to the reduction of plastic pollution. Additionally, it promotes the concept of a circular economy, which aims to minimise waste and maximise resource efficiency.

Despite their ubiquitous potential in technological waste management, certain challenges and considerations are associated with using nanotechnology. One of the major challenges in the field of nanotechnology is the translation of laboratory advances into industrial methods that are both scalable and cost-effective. The study of the environmental implications of nanoparticles, encompassing their potential toxicity and long-term behaviour in the environment, necessitates comprehensive assessment. Establishing regulatory frameworks is crucial in addressing the distinctive characteristics and potential hazards of nanomaterials to guarantee the safe and responsible use of nanotechnology in waste remediation.

7.8 SUMMARY

This chapter focuses on the exploration of advanced waste management technologies that are emerging in the field of sustainability. These technologies aim to efficiently address and reduce the environmental consequences associated with various types of technological waste, including metal extraction and mining, electronic, plastic, chemical, and nuclear waste. This discussion delves into modern waste management systems, scientific concepts and practical applications, offering comprehensive insights tailored for postgraduate students. The efficacy of these novel methodologies is substantiated by empirical case studies conducted in real-world settings, including diverse applications such as sensor-based ore sorting, bioleaching, and plasma cleanup. Furthermore, we explore the potential impact of nanotechnology on the sustainable management of waste. The chapter highlights the underlying concepts, advantages, drawbacks, and practical applications associated with these technologies. This thorough discourse provides a deep comprehension of the dynamic field of sustainable waste management, therefore facilitating the transition towards a more accountable and ecologically aware future for technological waste management and valorisation.

REFERENCES

Abdelbasir, Sabah M., Saad S. M. Hassan, Ayman H. Kamel, and Rania Seif El-Nasr. 2018. Status of electronic waste recycling techniques: A review. *Environmental Science and Pollution Research* 25 (17):16533–16547.

Agamuthu, P., K. M. Khidzir, and Fauziah Shahul Hamid. 2009. Drivers of sustainable waste management in Asia. *Waste Management & Research* 27 (7):625–633.

Aggelopoulos, Christos A. 2022. Recent advances of cold plasma technology for water and soil remediation: A critical review. *Chemical Engineering Journal* 428:131657.

Akash, S., Baskaran Sivaprakash, V. C. Vadivel Raja, Natarajan Rajamohan, and Govarthanan Muthusamy. 2022. Remediation techniques for uranium removal from polluted environment – Review on methods, mechanism and toxicology. *Environmental Pollution* 302:119068.

Akçalı, E., and S. Çetinkaya. 2011. Quantitative models for inventory and production planning in closed-loop supply chains. *International Journal of Production Research* 49 (8):2373–2407.

Ali, Sameh S., Tamer Elsamahy, Esraa A. Abdelkarim, et al. 2022. Biowastes for biodegradable bioplastics production and end-of-life scenarios in circular bioeconomy and biorefinery concept. *Bioresource Technology* 363:127869.

AliAkbari, Raouf, Yousef Marfavi, Elaheh Kowsari, and Seeram Ramakrishna. 2020. Recent studies on ionic liquids in metal recovery from E-waste and secondary sources by liquid-liquid extraction and electrodeposition: A review. *Materials Circular Economy* 2 (1):10.

Al-Sabagh, A. M., F. Z. Yehia, Gh Eshaq, A. M. Rabie, and A. E. ElMetwally. 2016. Greener routes for recycling of polyethylene terephthalate. *Egyptian Journal of Petroleum* 25 (1):53–64.

Alvarado-Hernández, L., G. T. Lapidus, and Federico González. 2019. Recovery of rare earths from waste cathode ray tube (CRT) phosphor powder with organic and inorganic ligands. *Waste Management* 95:53–58.

Araujo-Andrade, Cuauhtémoc, Elodie Bugnicourt, Laurent Philippet, et al. 2021. Review on the photonic techniques suitable for automatic monitoring of the composition of multi-materials wastes in view of their posterior recycling. *Waste Management & Research* 39 (5):631–651.

Arshad, Haroon, Shaharin A. Sulaiman, Zahid Hussain, Yasin Naz, and Firdaus Basrawi. 2017. Microwave assisted pyrolysis of plastic waste for production of fuels: A review. *MATEC Web Conference* 131, 02005.

Arya, Shashi, and Sunil Kumar. 2020. Bioleaching: Urban mining option to curb the menace of E-waste challenge. *Bioengineered* 11 (1):640–660.

Asante, Kwadwo Ansong, Yaw Amoyaw-Osei, and Tetsuro Agusa. 2019. E-waste recycling in Africa: Risks and opportunities. *Current Opinion in Green and Sustainable Chemistry* 18:109–117.

Aznar-Sánchez, José A., José J. García-Gómez, Juan F. Velasco-Muñoz, and Anselmo Carretero-Gómez. 2018. Mining waste and its sustainable management: Advances in worldwide research. *Minerals* 8 (7):284.

Babuponnusami, Arjunan, and Karuppan Muthukumar. 2014. A review on Fenton and improvements to the Fenton process for wastewater treatment. *Journal of Environmental Chemical Engineering* 2 (1):557–572.

Baniasadi, Mahsa, Farzane Vakilchap, Nazanin Bahaloo-Horeh, Seyyed Mohammad Mousavi, and Sebastien Farnaud. 2019. Advances in bioleaching as a sustainable method for metal recovery from e-waste: A review. *Journal of Industrial and Engineering Chemistry* 76:75–90.

Barrueto, Yahaira, Pía Hernández, Yecid P. Jiménez, and Jaime Morales. 2022. Properties and application of ionic liquids in leaching base/precious metals from e-waste. A review. *Hydrometallurgy* 212:105895.

Bastioli, Catia. 1998. Biodegradable materials — Present situation and future perspectives. *Macromolecular Symposia* 135 (1):193–204.

Bautista, P., A. F. Mohedano, J. A. Casas, J. A. Zazo, and J. J. Rodriguez. 2008. An overview of the application of Fenton oxidation to industrial wastewaters treatment. *Journal of Chemical Technology & Biotechnology* 83 (10):1323–1338.

Bernat, Katarzyna. 2023. Post-consumer plastic waste management: From collection and sortation to mechanical recycling. *Energies* 16 (8):3504.

Borja, Danilo, Kim A. Nguyen, Rene A. Silva, et al. 2016. Experiences and future challenges of bioleaching research in South Korea. *Minerals* 6 (4):128.

Bourg, Stéphane, Andreas Geist, Jean-Marc Adnet, Chris Rhodes, and Bruce C. Hanson. 2020. Partitioning and transmutation strategy R&D for nuclear spent fuel: the SACSESS and GENIORS projects. *EPJ Nuclear Sci. Technol.* 6, 2020.

Bu, Xiangning, January Kadenge Danstan, Ahmad Hassanzadeh, Ali Behrad Vakylabad, and Saeed Chehreh Chelgani. 2022. Metal extraction from ores and waste materials by ultrasound-assisted leaching-An overview. *Mineral Processing and Extractive Metallurgy Review* 45 (4):28–45.

Cacciuttolo, Carlos, and Alex Marinovic. 2023. Experiences of underground mine backfilling using mine tailings developed in the Andean Region of Peru: A green mining solution to reduce socio-environmental impacts. *Sustainability* 5 (17):12912.

Cacciuttolo, Carlos, and Fernando Valenzuela. 2022. Efficient use of water in tailings management: New technologies and environmental strategies for the future of mining. *Water* 14 (11):1741.

Capodaglio, Andrea G., Petr Hlavínek, and Massimo Raboni. 2015. Physico-chemical technologies for nitrogen removal from wastewaters: A review. *Revista ambiente agua* 10:481–498.

Chen, Jingwei, Tian Meng, Erwei Leng, and E. Jiaqiang. 2022. Review on metal dissolution characteristics and harmful metals recovery from electronic wastes by supercritical water. *Journal of Hazardous Materials* 424:127693.

Comninellis, Christos, Agnieszka Kapalka, Sixto Malato, Simon A. Parsons, Ioannis Poulios, and Dionissios Mantzavinos. 2008. Advanced oxidation processes for water treatment: Advances and trends for R&D. *Journal of Chemical Technology & Biotechnology* 83 (6):769–776.

Cui, Jirang, and Eric Forssberg. 2003. Mechanical recycling of waste electric and electronic equipment: A review. *Journal of Hazardous Materials* 99 (3):243–263.

Cui, Jirang, and Lifeng Zhang. 2008. Metallurgical recovery of metals from electronic waste: A review. *Journal of Hazardous Materials* 158 (2):228–256.

Deng, Yang, and Renzun Zhao. 2015. Advanced oxidation processes (AOPs) in wastewater treatment. *Current Pollution Reports* 1 (3):167–176.

Diez-Olivan, Alberto, Javier Del Ser, Diego Galar, and Basilio Sierra. 2019. Data fusion and machine learning for industrial prognosis: Trends and perspectives towards Industry 4.0. *Information Fusion* 50:92–111.

Dimitriadis, Athanasios, and Stella Bezergianni. 2017. Hydrothermal liquefaction of various biomass and waste feedstocks for biocrude production: A state of the art review. *Renewable and Sustainable Energy Reviews* 68:113–125.

Doiphode, Shubham, J. Dhanuskar, and A. R. Deshmukh. A review on use of gold mine waste in preparation of concrete. International Journal for Research in Applied Science and Engineering Technology. 11:595–598.

Driussi, Catherine, and Janis Jansz. 2006. Technological options for waste minimisation in the mining industry. *Journal of Cleaner Production* 14 (8):682–688.

Durand, J. F. 2012. The impact of gold mining on the Witwatersrand on the rivers and karst system of Gauteng and North West Province, South Africa. *Journal of African Earth Sciences* 68:24–43.

Dutta, Tanushree, Ki-Hyun Kim, Akash Deep, et al. 2018. Recovery of nanomaterials from battery and electronic wastes: A new paradigm of environmental waste management. *Renewable and Sustainable Energy Reviews* 82:3694–3704.

Edraki, Mansour, Thomas Baumgartl, Emmanuel Manlapig, Dee Bradshaw, Daniel M. Franks, and Chris J. Moran. 2014. Designing mine tailings for better environmental, social and economic outcomes: A review of alternative approaches. *Journal of Cleaner Production* 84:411–420.

Evode, Niyitanga, Sarmad Ahmad Qamar, Muhammad Bilal, Damià Barceló, and Hafiz M. N. Iqbal. 2021. Plastic waste and its management strategies for environmental sustainability. *Case Studies in Chemical and Environmental Engineering* 4:100142.

Faris, Nebeal, Mark I. Pownceby, Warren J. Bruckard, and Miao Chen. 2023. The Direct leaching of nickel sulfide flotation concentrates – A historic and state-of-the-art review Part I: Piloted processes and commercial operations. *Mineral Processing and Extractive Metallurgy Review* 44 (6):407–435.

Forti, Vanessa, Cornelis P Balde, Ruediger Kuehr, and Garam Bel. 2020. "The Global E-waste Monitor 2020: Quantities, flows and the circular economy potential." (2020).

Ghimire, Hem, and Parisa A. Ariya. 2020. E-wastes: Bridging the knowledge gaps in global production budgets, composition, recycling and sustainability implications. *Sustainable Chemistry* 1 (2):154–182.

Giurco, Damien, Benjamin McLellan, Daniel M. Franks, Keisuke Nansai, and Timothy Prior. 2014. Responsible mineral and energy futures: Views at the nexus. *Journal of Cleaner Production* 84:322–338.

Gomez, E., D. Amutha Rani, C. R. Cheeseman, D. Deegan, M. Wise, and A. R. Boccaccini. 2009. Thermal plasma technology for the treatment of wastes: A critical review. *Journal of Hazardous Materials* 161 (2):614–626.

Goto, Motonobu. 2009. Chemical recycling of plastics using sub- and supercritical fluids. *The Journal of Supercritical Fluids* 47 (3):500–507.

Hao, Mengjie, Yanfang Liu, Weijin Wu, et al. 2023. Advanced porous adsorbents for radionuclides elimination. *EnergyChem* 5 (4):100101.

Hu, Yulin, Mai Attia, Emir Tsabet, Ahmad Mohaddespour, Muhammad Tajammal Munir, and Sherif Farag. 2021. Valorization of waste tire by pyrolysis and hydrothermal liquefaction: A mini-review. *Journal of Material Cycles and Waste Management* 23 (5):1737–1750.

Huang, Jan-Chan, Aditya S. Shetty, and Ming-Song Wang. 1990. Biodegradable plastics: A review. *Advances in Polymer Technology* 10 (1):23–30.

Ippolito, Nicolò Maria, Francesco Ferella, Valentina Innocenzi, et al. 2021. Effect of mechanical activation on terbium dissolution from waste fluorescent powders. *Minerals Engineering* 167:106906.

Ippolito, Nicolò Maria, Valentina Innocenzi, Ida De Michelis, Franco Medici, and Francesco Vegliò. 2017. Rare earth elements recovery from fluorescent lamps: A new thermal pretreatment to improve the efficiency of the hydrometallurgical process. *Journal of Cleaner Production* 153:287–298.

Islam, Aminul, Tofayal Ahmed, Md Rabiul Awual, et al. 2020. Advances in sustainable approaches to recover metals from e-waste-A review. *Journal of Cleaner Production* 244:118815.

Islam, Aminul, A. M. Swaraz, Siow Hwa Teo, et al. 2021. Advances in physiochemical and biotechnological approaches for sustainable metal recovery from e-waste: A critical review. *Journal of Cleaner Production* 323:129015.

Islam, Md Tasbirul, and Nazmul Huda. 2018. Reverse logistics and closed-loop supply chain of Waste Electrical and Electronic Equipment (WEEE)/E-waste: A comprehensive literature review. *Resources, Conservation and Recycling* 137:48–75.

Iwamoto, Tomotada, and Masao Nasu. 2001. Current bioremediation practice and perspective. *Journal of Bioscience and Bioengineering* 92 (1):1–8.

Jain, Keerti, Anand S. Patel, Vishwas P. Pardhi, and Swaran J. Flora. 2021. Nanotechnology in wastewater management: A new paradigm towards wastewater treatment. *Molecules* 26 (6):1797.

Jiang, Cheng-chun, and Jia-fa Zhang. 2007. Progress and prospect in electro-Fenton process for wastewater treatment. *Journal of Zhejiang University-SCIENCE A* 8 (7):1118–1125.

Johansson, Christian, Jenny Elfsberg, Tobias C. Larsson, et al. 2016. Urban mining as A case for PSS. *Procedia CIRP* 47:460–465.

Kalisz, Szymon, Katarzyna Kibort, Joanna Mioduska, Marek Lieder, and Aleksandra Małachowska. 2022. Waste management in the mining industry of metals ores, coal, oil and natural gas – A review. *Journal of Environmental Management* 304:114239.

Kalpakli, Ahmet Orkun, Mine Caymaz, Sedat Ilhan, and Mert Zoraga. 2021. Recovery of Zn as ZnO from steelmaking waste materials by mechanochemical leaching, solvent extraction, precipitation, and thermal decomposition route. *Journal of Sustainable Metallurgy* 7 (1):277–290.

Kaushal, Jyoti, Madhu Khatri, and Shailendra Kumar Arya. 2021. Recent insight into enzymatic degradation of plastics prevalent in the environment: A mini – review. *Cleaner Engineering and Technology* 2:100083.

Khalidy, Reza, and Rafael M. Santos. 2021. The fate of atmospheric carbon sequestrated through weathering in mine tailings. *Minerals Engineering* 163:106767.

Khan, Shamshad, Mu Naushad, Adel Al-Gheethi, and Jibran Iqbal. 2021. Engineered nanoparticles for removal of pollutants from wastewater: Current status and future prospects of nanotechnology for remediation strategies. *Journal of Environmental Chemical Engineering* 9 (5):106160.

Kiddee, Peeranart, Ravi Naidu, and Ming H. Wong. 2013. Electronic waste management approaches: An overview. *Waste Management* 33 (5):1237–1250.

Kiran, Raj, Catalin Teodoriu, Younas Dadmohammadi, et al. 2017. Identification and evaluation of well integrity and causes of failure of well integrity barriers (A review). *Journal of Natural Gas Science and Engineering* 45:511–526.

Kothari, Richa, V. V. Tyagi, and Ashish Pathak. 2010. Waste-to-energy: A way from renewable energy sources to sustainable development. *Renewable and Sustainable Energy Reviews* 14 (9):3164–3170.

Krishnan, Santhana, Nor Syahidah Zulkapli, Hesam Kamyab, et al. 2021. Current technologies for recovery of metals from industrial wastes: An overview. *Environmental Technology & Innovation* 22:101525.

Kumar, Amit, Gaurav Sharma, Mu Naushad, et al. 2020. Bio-inspired and biomaterials-based hybrid photocatalysts for environmental detoxification: A review. *Chemical Engineering Journal* 382:122937.

Kumari, Rima, and Sukha Ranjan Samadder. 2022. A critical review of the pre-processing and metals recovery methods from e-wastes. *Journal of Environmental Management* 320:115887.

Kurniawan, Tonni Agustiono, Wai-hung Lo, and G. Y. S. Chan. 2006. Radicals-catalyzed oxidation reactions for degradation of recalcitrant compounds from landfill leachate. *Chemical Engineering Journal* 125 (1):35–57.

Kurniawan, Tonni Agustiono, Mohd Hafiz Dzarfan Othman, Deepak Singh, et al. 2022. Technological solutions for long-term storage of partially used nuclear waste: A critical review. *Annals of Nuclear Energy* 166:108736.

Kuzmanović, Maja, Laurens Delva, Ludwig Cardon, and Kim Ragaert. 2021. The feasibility of using the MFC concept to upcycle mixed recycled plastics. *Sustainability* 13 (2):689.

La Fuente, Carla Ivonne Arias, Bianca Chieregato Maniglia, and Carmen Cecília Tadini. 2023. Biodegradable polymers: A review about biodegradation and its implications and applications. *Packaging Technology and Science* 36 (2):81–95.

Lagaron, Jose M., and Amparo Lopez-Rubio. 2011. Nanotechnology for bioplastics: Opportunities, challenges and strategies. *Trends in Food Science & Technology* 22 (11):611–617.

Laredo, Georgina C., Joel Reza, and Edith Meneses Ruiz. 2023. Hydrothermal liquefaction processes for plastics recycling: A review. *Cleaner Chemical Engineering* 5:100094.

Lepawsky, Josh. 2015. The changing geography of global trade in electronic discards: Time to rethink the e-waste problem. *The Geographical Journal* 181 (2):147–159.

Levchenko, Igor, Mohandas Mandhakini, Karthika Prasad, et al. 2022. Functional nanomaterials from waste and low-value natural products: A technological approach level. *Advanced Materials Technologies* 7 (11):2101471.

Li, J., M. Barwood, and S. Rahimifard. 2018. Robotic disassembly for increased recovery of strategically important materials from electrical vehicles. *Robotics and Computer-Integrated Manufacturing* 50:203–212.

Li, Jie, Xiangxue Wang, Guixia Zhao, et al. 2018. Metal–organic framework-based materials: Superior adsorbents for the capture of toxic and radioactive metal ions. *Chemical Society Reviews* 47 (7):2322–2356.

Li, Kuo, and Zhenming Xu. 2019. A review of current progress of supercritical fluid technologies for e-waste treatment. *Journal of Cleaner Production* 227:794–809.

Li, Ning, Hengxin Liu, Zhanjun Cheng, Beibei Yan, Guanyi Chen, and Shaobin Wang. 2022. Conversion of plastic waste into fuels: A critical review. *Journal of Hazardous Materials* 424:127460.

Li, Xin, Pengfei Xing, Xinghong Du, Shuaibo Gao, and Chen Chen. 2017. Influencing factors and kinetics analysis on the leaching of iron from boron carbide waste-scrap with ultrasound-assisted method. *Ultrasonics Sonochemistry* 38:84–91.

Li, Ying, Jinhui Li, Shusheng Chen, and Weihua Diao. 2012. Establishing indices for groundwater contamination risk assessment in the vicinity of hazardous waste landfills in China. *Environmental Pollution* 165:77–90.

Liao, Chunfa, Zhenyuan Li, Yanliang Zeng, Jingyuan Chen, Liqin Zhong, and Li Wang. 2017. Selective extraction and recovery of rare earth metals from waste fluorescent powder using alkaline roasting-leaching process. *Journal of Rare Earths* 35 (10):1008–1013.

Liu, Bowen, Nengwu Zhu, Yao Li, Pingxiao Wu, Zhi Dang, and Yixin Ke. 2019. Efficient recovery of rare earth elements from discarded NdFeB magnets. *Process Safety and Environmental Protection* 124:317–325.

Liu, Qingsheng, Tao Tu, Hao Guo, Huajin Cheng, and Xuezhong Wang. 2021. High-efficiency simultaneous extraction of rare earth elements and iron from NdFeB waste by oxalic acid leaching. *Journal of Rare Earths* 39 (3):323–330.

Luo, Xianping, Kunzhong He, Yan Zhang, Pengyu He, and Yongbing Zhang. 2022. A review of intelligent ore sorting technology and equipment development. *International Journal of Minerals, Metallurgy and Materials* 29 (9):1647–1655.

Luque-García, J. L., and M. D. Luque de Castro. 2003. Ultrasound: A powerful tool for leaching. *TrAC Trends in Analytical Chemistry* 22 (1):41–47.

Mahore, Samit, R. Y. Mahore, and Sanjay J. Dhoble. 2023. 10 – Economics of electronic waste (e-waste): A critical review of negative externalities and the circular economy. In *360-Degree Waste Management, Volume 1*, edited by N. A. Raut, D. M. Kokare, B. A. Bhanvase, K. R. Randive, and S. J. Dhoble: Elsevier.

Manos, Manolis J., and Mercouri G. Kanatzidis. 2016. Metal sulfide ion exchangers: Superior sorbents for the capture of toxic and nuclear waste-related metal ions. *Chemical Science* 7 (8):4804–4824.

Marafi, M., and A. Stanislaus. 2008. Spent hydroprocessing catalyst management: A review: Part II. Advances in metal recovery and safe disposal methods. *Resources, Conservation and Recycling* 53 (1):1–26.

M'Arimi, M. M., C. A. Mecha, A. K. Kiprop, and R. Ramkat. 2020. Recent trends in applications of advanced oxidation processes (AOPs) in bioenergy production: Review. *Renewable and Sustainable Energy Reviews* 121:109669.

Marrone, Philip A. 2013. Supercritical water oxidation—Current status of full-scale commercial activity for waste destruction. *The Journal of Supercritical Fluids* 79:283–288.

Marsden, J. O., and M. M. Botz. 2017. Heap leach modeling — A review of approaches to metal production forecasting. *Minerals & Metallurgical Processing* 34 (2):53–64.

Martins, Natalia P., Sumit Srivastava, Francisco V. Simão, et al. 2021. Exploring the potential for utilization of medium and highly sulfidic mine tailings in construction materials: A review. *Sustainability* 13 (21):121150.

Matinde, E., G. S. Simate, and S. Ndlovu. 2018. Mining and metallurgical wastes: A review of recycling and re-use practices. *Journal of the Southern African Institute of Mining and Metallurgy* 118:825–844.

Medfu Tarekegn, Molalign, Fikirte Zewdu Salilih, and Alemitu Iniyehu Ishetu. 2020. Microbes used as a tool for bioremediation of heavy metal from the environment. *Cogent Food & Agriculture* 6 (1):1783174.

Menka, Nazune, Rob Root, and Jon Chorover. 2014. Bioaccessibility, release kinetics, and molecular speciation of arsenic and lead in geo-dusts from the Iron King Mine Federal Superfund site in Humboldt, Arizona. *Reviews on Environmental Health,* 29 (1–2):23–27.

Miandad, R., M. A. Barakat, Asad S. Aburiazaiza, M. Rehan, and A. S. Nizami. 2016. Catalytic pyrolysis of plastic waste: A review. *Process Safety and Environmental Protection* 102:822–838.

Mikula, Katarzyna, Dawid Skrzypczak, Grzegorz Izydorczyk, et al. 2021. 3D printing filament as a second life of waste plastics—A review. *Environmental Science and Pollution Research* 28 (10):12321–12333.

Minor, Ann-Joelle, Ruben Goldhahn, Liisa Rihko-Struckmann, and Kai Sundmacher. 2023. Chemical recycling processes of Nylon 6 to Caprolactam: Review and techno-economic assessment. *Chemical Engineering Journal* 474:145333.

Misra, Virendra, and S. D. Pandey. 2005. Hazardous waste, impact on health and environment for development of better waste management strategies in future in India. *Environment International* 31 (3):417–431.

Modise, E. G., A. M. Zungeru, B. Mtengi, and A. U. Ude. 2022. Sensor-based ore sorting—A review of current use of electromagnetic spectrum in sorting. *IEEE Access* 10:112307–112326.

Mohajerani, Abbas, David Suter, Tristan Jeffrey-Bailey, et al. 2019. Recycling waste materials in geopolymer concrete. *Clean Technologies and Environmental Policy* 21 (3):493–515.

Mohanan, Nisha, Zahra Montazer, Parveen K. Sharma, and David B. Levin. 2020. Microbial and enzymatic degradation of synthetic plastics. *Frontiers in Microbiology* 11:1–22.

Monsigny, Louis, Jean-Claude Berthet, and Thibault Cantat. 2018. Depolymerization of waste plastics to monomers and chemicals using a hydrosilylation strategy facilitated by Brookhart's Iridium(III) catalyst. *ACS Sustainable Chemistry & Engineering* 6 (8):10481–10488.

Munir, M. T., I. Mardon, S. Al-Zuhair, A. Shawabkeh, and N. U. Saqib. 2019. Plasma gasification of municipal solid waste for waste-to-value processing. *Renewable and Sustainable Energy Reviews* 116:109461.

Murthy, Venkatesha, and Seeram Ramakrishna. 2022. A review on global E-waste management: Urban mining towards a sustainable future and circular economy. *Sustainability* 14 (2):647.

Ni'am, Achmad Chusnun, Ya-Fen Wang, Shyh-Wei Chen, and Sheng-Jie You. 2019. Recovery of rare earth elements from waste permanent magnet (WPMs) via selective leaching using the Taguchi method. *Journal of the Taiwan Institute of Chemical Engineers* 97:137–145.

Ni'am, Achmad Chusnun, Ya-Fen Wang, Shyh-Wei Chen, Gen-Mu Chang, and Sheng-Jie You. 2020. Simultaneous recovery of rare earth elements from waste permanent magnets (WPMs) leach liquor by solvent extraction and hollow fiber supported liquid membrane. *Chemical Engineering and Processing – Process Intensification* 148:107831.

Nkuna, Rosina, Grace N. Ijoma, Tonderayi S. Matambo, and Ngonidzashe Chimwani. 2022. Accessing metals from low-grade ores and the environmental impact considerations: A review of the perspectives of conventional versus bioleaching strategies. *Minerals* 12 (5):506.

Nyakuma, Bemgba B., and Tertsegha J. P. Ivase. 2021. Emerging trends in sustainable treatment and valorisation technologies for plastic wastes in Nigeria: A concise review. *Environmental Progress & Sustainable Energy* 40 (5):e13660.

Odebiyi, Oluwasegun Samuel, Hao Du, Biao Liu, and Shaona Wang. 2022. Sustainability of valuable metals recovery from hazardous industrial solid wastes: The role of mechanical activation. *Journal of Sustainable Metallurgy* 8 (4):1393–1421.

Olson, G. J., J. A. Brierley, and C. L. Brierley. 2003. Bioleaching review part B. *Applied Microbiology and Biotechnology* 63 (3):249–257.

O'Neill Marie, S., Michael Jerrett, Ichiro Kawachi, et al. 2003. Health, wealth, and air pollution: Advancing theory and methods. *Environmental Health Perspectives* 111 (16):1861–1870.

Ou, Zhiyuan, Jinhui Li, and Zhishi Wang. 2015. Application of mechanochemistry to metal recovery from second-hand resources: A technical overview. *Environmental Science: Processes & Impacts* 17 (9):1522–1530.

Pabby, A. K., B. Swain, N. L. Sonar, et al. 2022. Radioactive waste processing using membranes: State of the art technology, challenges and perspectives. *Separation & Purification Reviews* 51 (2):143–173.

Panda, Achyut K., R. K. Singh, and D. K. Mishra. 2010. Thermolysis of waste plastics to liquid fuel: A suitable method for plastic waste management and manufacture of value added products—A world prospective. *Renewable and Sustainable Energy Reviews* 14 (1):233–248.

Pant, Deepak, Deepika Joshi, Manoj K. Upreti, and Ravindra K. Kotnala. 2012. Chemical and biological extraction of metals present in E waste: A hybrid technology. *Waste Management* 32 (5):979–990.

Park, Sebeom, and Yosoon Choi. 2020. Applications of unmanned aerial vehicles in mining from exploration to reclamation: A review. *Minerals* 10 (8):663.

Patil, Ajay B., Mohamed Tarik, Rudolf P. W. J. Struis, and Christian Ludwig. 2021. Exploiting end-of-life lamps fluorescent powder e-waste as a secondary resource for critical rare earth metals. *Resources, Conservation and Recycling* 164:105153.

Pavón, S., T. Lorenz, A. Fortuny, A. M. Sastre, and M. Bertau. 2021. Rare earth elements recovery from secondary wastes by solid-state chlorination and selective organic leaching. *Waste Management* 122:55–63.

Peng, Yujie, Yunpu Wang, Linyao Ke, et al. 2022. A review on catalytic pyrolysis of plastic wastes to high-value products. *Energy Conversion and Management* 254:115243.

Phillip, Esther, Thye F. Choo, Nurul W. Khairuddin, and Rehab O. Abdel Rahman. 2023. On the sustainable utilization of geopolymers for safe management of radioactive waste: A review. *Sustainability* 15 (2):1117.

Preetam, Amrita, Prashant Ram Jadhao, S. N. Naik, K. K. Pant, and Vivek Kumar. 2023. Supercritical fluid technology–An eco-friendly approach for resource recovery from e-waste and plastic waste: A review. *Separation and Purification Technology* 304:122314.

Rahimi, AliReza, and Jeannette M. García. 2017. Chemical recycling of waste plastics for new materials production. *Nature Reviews Chemistry* 1 (6):0046.

Rahman, R. O. Abdel, H. A. Ibrahium, and Yung-Tse Hung. 2011. Liquid radioactive wastes treatment: A review. *Water* 3 (2):551–565.

Ramírez-García, Robert, Nisarg Gohil, and Vijai Singh. 2019. Chapter 21 – Recent advances, challenges, and opportunities in bioremediation of hazardous materials. In *Phytomanagement of Polluted Sites*, edited by V. C. Pandey, and K. Bauddh: Elsevier.

Ramprasad, C., Willis Gwenzi, Nhamo Chaukura, et al. 2022. Strategies and options for the sustainable recovery of rare earth elements from electrical and electronic waste. *Chemical Engineering Journal* 442:135992.

Rasoulnia, Payam, Robert Barthen, Jaakko A. Puhakka, and Aino-Maija Lakaniemi. 2021. Leaching of rare earth elements and base metals from spent NiMH batteries using gluconate and its potential bio-oxidation products. *Journal of Hazardous Materials* 414:125564.

Rautela, Rahul, Shashi Arya, Shilpa Vishwakarma, Jechan Lee, Ki-Hyun Kim, and Sunil Kumar. 2021. E-waste management and its effects on the environment and human health. *Science of The Total Environment* 773:145623.

Resende, Luciene V., and Carlos A. Morais. 2010. Study of the recovery of rare earth elements from computer monitor scraps – Leaching experiments. *Minerals Engineering* 23 (3):277–280.

Riley, Brian J., John D. Vienna, Denis M. Strachan, John S. McCloy, and James L. Jerden. 2016. Materials and processes for the effective capture and immobilization of radioiodine: A review. *Journal of Nuclear Materials* 470:307–326.

Robben, Christopher, and Hermann Wotruba. 2019. Sensor-based ore sorting technology in mining—Past, present and future. *Minerals* 9(9), 523.

Roohi, Kulsoom Bano, Mohammed Kuddus, et al. 2017. Microbial enzymatic degradation of biodegradable plastics. *Current Pharmaceutical Biotechnology* 18 (5):429–440.

Rueda-Marquez, Juan José, Irina Levchuk, Pilar Fernández Ibañez, and Mika Sillanpää. 2020. A critical review on application of photocatalysis for toxicity reduction of real wastewaters. *Journal of Cleaner Production* 258:120694.

Sajjad, Wasim, Guodong Zheng, Ghufranud Din, Xiangxian Ma, Muhammad Rafiq, and Wang Xu. 2019. Metals extraction from sulfide ores with microorganisms: The bioleaching technology and recent developments. *Transactions of the Indian Institute of Metals* 72 (3):559–579.

Salvatores, M., and G. Palmiotti. 2011. Radioactive waste partitioning and transmutation within advanced fuel cycles: Achievements and challenges. *Progress in Particle and Nuclear Physics* 66 (1):144–166.

Salwadi, Yadaiah, and Sathish Kumar Kurapati. 2022. The role of nanomaterials in radioactive waste reduction and decontamination: A concise review. *Paper Read at International Conference on Nanotechnology for Sustainable Living and Environment,* Springer Nature Singapore, 141–157.

Sanito, Raynard Christianson, Sheng-Jie You, and Ya-Fen Wang. 2021. Application of plasma technology for treating e-waste: A review. *Journal of Environmental Management* 288:112380.

Sanlisoy, A., and M. O. Carpinlioglu. 2017. A review on plasma gasification for solid waste disposal. *International Journal of Hydrogen Energy* 42 (2):1361–1365.

Sarc, R., A. Curtis, L. Kandlbauer, K. Khodier, K. E. Lorber, and R. Pomberger. 2019. Digitalisation and intelligent robotics in value chain of circular economy oriented waste management – A review. *Waste Management* 95:476–492.

Schmieder, H., and J. Abeln. 1999. Supercritical water oxidation: State of the art. *Chemical Engineering & Technology* 22 (11):903–908.

Schyns, Zoé O. G., and Michael P. Shaver. 2021. Mechanical recycling of packaging plastics: A review. *Macromolecular Rapid Communications* 42 (3):2000415.

Sengupta, Diyasha, I. M. S. K. Ilankoon, Kai Dean Kang, and Meng Nan Chong. 2022. Circular economy and household e-waste management in India: Integration of formal and informal sectors. *Minerals Engineering* 184:107661.

Serag, Eman. 2022. The challenges and opportunities of the transformation of e-waste into value-added materials: e-waste management approach. *Environmental Sustainability* 5 (3):277–288.

Sharma, Manu, Sudhanshu Joshi, and Kannan Govindan. 2021. Issues and solutions of electronic waste urban mining for circular economy transition: An Indian context. *Journal of Environmental Management* 290:112373.

Sherwood, James. 2020. Closed-loop recycling of polymers using solvents: Remaking plastics for a circular economy. *Johnson Matthey Technology Review* 64 (1):4–15.

Sillitoe, Richard H., and John F. H. Thompson. 2005. Changes in mineral exploration practice: Consequences for discovery. In *Wealth Creation in the Minerals Industry: Integrating Science, Business, and Education*, edited by M. D. Doggett, and J. R. Parry: Society of Economic Geologists.

Singh, Pardeep, Rajat Jain, Neha Srivastava, et al. 2017. Current and emerging trends in bioremediation of petrochemical waste: A review. *Critical Reviews in Environmental Science and Technology* 47 (3):155–201.

Soh, S. L., S. K. Ong, and A. Y. C. Nee. 2014. Design for disassembly for remanufacturing: Methodology and technology. *Procedia CIRP* 15:407–412.

Song, Wen, Wei Song, Haihong Gu, and Fuping Li. 2020. Progress in the remote sensing monitoring of the ecological environment in mining areas. *International Journal of Environmental Research and Public Health* 17 (6):1846.

Su, Xiang, Wenqi Xie, and Xiaoqi Sun. 2021. A sustainable [P6,6,6,14]2[OPBOA]-based separation process of rare earth and transition metal in waste NiMH battery. *Minerals Engineering* 160:106641.

Sum, Elaine Y. L. 1991. The recovery of metals from electronic scrap. *JOM* 43 (4):53–61.

Tan, Quanyin, Chao Deng, and Jinhui Li. 2017. Enhanced recovery of rare earth elements from waste phosphors by mechanical activation. *Journal of Cleaner Production* 142:2187–2191.

Tan, Quanyin, and Jinhui Li. 2015. Recycling metals from wastes: A novel application of mechanochemistry. *Environmental Science & Technology* 49 (10):5849–5861.

Tang, XingYing, YouChang Zheng, ZeQin Liao, YingHui Wang, JianQiao Yang, and Jianjun Cai. 2021. A review of developments in process flow for supercritical water oxidation. *Chemical Engineering Communications* 208 (10):1494–1510.

Thiounn, Timmy, and Rhett C. Smith. 2020. Advances and approaches for chemical recycling of plastic waste. *Journal of Polymer Science* 58 (10):1347–1364.

Trojanowicz, Marek, Kamila Kołacińska, and Jay W. Grate. 2018. A review of flow analysis methods for determination of radionuclides in nuclear wastes and nuclear reactor coolants. *Talanta* 183:70–82.

Tutton, Carl G., Steven B. Young, and Komal Habib. 2022. Pre-processing of e-waste in Canada: Case of a facility responding to changing material composition. *Resources, Environment and Sustainability* 9:100069.

Udayakumar, Sanjith, Muhammad Irfan Bin Abd Razak, and Suhaina Ismail. 2022. Recovering valuable metals from Waste Printed Circuit Boards (WPCB): A short review. *Materials Today: Proceedings* 66:3062–3070.

Valerio, Oscar, Rajendran Muthuraj, and Amandine Codou. 2020. Strategies for polymer to polymer recycling from waste: Current trends and opportunities for improving the circular economy of polymers in South America. *Current Opinion in Green and Sustainable Chemistry* 25:100381.

Villegas, Laura G. Cordova, Neda Mashhadi, Miao Chen, Debjani Mukherjee, Keith E. Taylor, and Nihar Biswas. 2016. A short review of techniques for phenol removal from wastewater. *Current Pollution Reports* 2 (3):157–167.

Wang, X., P. Wang, X. Zhang, Y. Wan, H. Shi, and W. Liu. 2022. Target electromagnetic detection method in underground environment: A review. *IEEE Sensors Journal* 22 (14):13835–13852.

Xavier, Lúcia Helena, Ellen Cristine Giese, Ana Cristina Ribeiro-Duthie, and Fernando Antonio Freitas Lins. 2021. Sustainability and the circular economy: A theoretical approach focused on e-waste urban mining. *Resources Policy* 74:101467.

Xavier, Lúcia Helena, Marianna Ottoni, and Leonardo Picanço Peixoto Abreu. 2023. A comprehensive review of urban mining and the value recovery from e-waste materials. *Resources, Conservation and Recycling* 190:106840.

Xavier, Lúcia Helena, Marianna Ottoni, and Josh Lepawsky. 2021. Circular economy and e-waste management in the Americas: Brazilian and Canadian frameworks. *Journal of Cleaner Production* 297:126570.

Xie, Guangyu, Xinjiang Hu, Yuxuan Du, et al. 2021. Light-driven breakdown of microcystin-LR in water: A critical review. *Chemical Engineering Journal* 417:129244.

Xu, Da-Mao, Chang-Lin Zhan, Hong-Xia Liu, and Han-Zhi Lin. 2019. A critical review on environmental implications, recycling strategies, and ecological remediation for mine tailings. *Environmental Science and Pollution Research* 26 (35):35657–35669.

Yaashikaa, P. R., B. Priyanka, P. Senthil Kumar, S. Karishma, S. Jeevanantham, and Sravya Indraganti. 2022. A review on recent advancements in recovery of valuable and toxic metals from e-waste using bioleaching approach. *Chemosphere* 287:132230.

Yang, Hee-Man, Chan Woo Park, and Kune-Woo Lee. 2018. Polymeric coatings for surface decontamination and ecofriendly volume reduction of radioactive waste after use. *Progress in Nuclear Energy* 104:67–74.

Yang, Xiong, Lihu Liu, Wenfeng Tan, Chengshuai Liu, Zhi Dang, and Guohong Qiu. 2020. Remediation of heavy metal contaminated soils by organic acid extraction and electrochemical adsorption. *Environmental Pollution* 264:114745.

Yin, Xiaofei, Xiangmiao Tian, Yufeng Wu, et al. 2018. Recycling rare earth elements from waste cathode ray tube phosphors: Experimental study and mechanism analysis. *Journal of Cleaner Production* 205:58–66.

Yudaev, Pavel A., and Evgeniy M. Chistyakov. 2022. Ionic liquids as components of systems for metal extraction. *ChemEngineering* 6 (1):6.

Yurramendi, Lourdes, Lukas Gijsemans, Federica Forte, Jose Luis Aldana, Carmen del Río, and Koen Binnemans. 2019. Enhancing rare-earth recovery from lamp phosphor waste. *Hydrometallurgy* 187:38–44.

Zeng, Tao, Guohua Chen, Yunfeng Yang, Genserik Reniers, Yixin Zhao, and Xia Liu. 2020. A systematic literature review on safety research related to chemical industrial parks. *Sustainability* 12 (14):5753.

Zhang, Hao, Danyan Ma, Rongliang Qiu, Yetao Tang, and Changming Du. 2017. Non-thermal plasma technology for organic contaminated soil remediation: A review. *Chemical Engineering Journal* 313:157–170.

Zhi, Hailan, Shuainan Ni, Xiang Su, Wenqi Xie, Hepeng Zhang, and Xiaoqi Sun. 2022. Separation and recovery of rare earth from waste nickel-metal hydride batteries by phosphate based extraction-precipitation. *Journal of Rare Earths* 40 (6):974–980.

Zhou, Nan, Leilei Dai, Yuancai Lv, et al. 2021. Catalytic pyrolysis of plastic wastes in a continuous microwave assisted pyrolysis system for fuel production. *Chemical Engineering Journal* 418:129412.

8 Summary and Future Direction

Krishanu Biswas

8.1 NEED FOR TECHNOLOGICAL WASTE MANAGEMENT

We generate a humongous amount of waste in the world today. Trends show the generation of waste has shown a positive slope [1]. If this trend continues for the next 50–100 years, the world will no longer be a place to live. The linear economic model for economic growth for the last 250 years has led to generation waste, waste, and waste. This has also led to significant damage to strategic and natural resources. Therefore, the transition to a circular economy with zero wastage is considered to be the best revolution. This also provides the best economic opportunity. The circular economy represents a massive opportunity for us to create a distinct advantage over the linear economy. This is also termed as circular advantage. It is a basically a disruptive way to produce as well as consume by using innovative technologies and business models. The detailed resources reveal that the amount of waste generated by 2030 will be so large that the reward for turning this waste to wealth will be about $4.5 trillion. The waste, what we conceive now, is a massive amount of underutilized natural resources. The circular economy will make sure everything in this world has a value- there is nothing like a waste.

8.2 CURRENT SCENARIO OF TECHNOLOGICAL WASTE MANAGEMENT AND ADVENT OF SUSTAINABLE WASTE MANAGEMENT TECHNOLOGIES

The detailed discussion in the last seven chapters has categorically indicated that current scenario of waste management in the world is dismally unsatisfactory. The current business model (linear economics model) is tantamount to "business as usual"; this is what leading to major trouble for waste management in the world. We need to understand that mother Earth cannot be considered as a "place to plunder all and dispose all after use". The resources are limited on the Earth. In fact, some of non-renewable resources (oil, gas, and strategic metals) have fast been dwindling. This is due to pervasive mining and rough and unmindful way of utilization of resources. Similarly the other renewable resources (water, air, greenery, some other metals etc.) are technically renewable but heavily stressed in various parts of the world. If we continue using the linear model of economic development, we will exhaust the resources

DOI: 10.1201/9781003619680-8

FIGURE 8.1 A cartoon showing that mounting waste requires us to rebooting the mother Earth. (From https://cartoonstock.com/)

available on the mother Earth too soon to the breaking point. It is evident massive population growth and its future trend are making the demand on the resources too high for the system to provide any meaningful solution. Added to our insensible behavior, which treats the impact of mounting waste as the problem of "someone else" is making the landfills, water bodies and space as junkyard for waste. Therefore, the current situation in alarming and akin to the cartoon shown in Figure 8.1.

The waste management in both developed and developing countries is not as rosy as it is projected. The wastes, depending on type and nature, need to be collected and treated in such a way that we do not create more waste. We need to understand that the treatment of waste must not lead to generation of further waste. Rather we must recover and recycle various wastes in a practical manner. The municipal industrial, and contruction waste, occupying the lion's share need to be treated differently using deferent innovation technological solution. However, very few counties (Sweden, Denmark are notable examples) have developed and implemented one of the best technological waste management systems, known as zero-waste management. In the beginning of the current century, the prices of various commodities significantly rise and the trend is highly positive. This is significantly due to scarce availability of resources, albeit geopolitical factors. In reality, the multinational companies are not finding enough resources to meet for growth. The majority of world population (50% = 3.5 billion people) just live on US$ 3 per day and 75% of world pollution (~5 billion) can only afford US$10 per day [2]. Therefore, the future growth model must involve 5 billion (or more in future) to share the resources available in the world. This means we need radical change of resource generation and management. Definitely, the sustainable future needs embracing the circular economy. This concept has widely been deliberated for several decades. At the core ot it this categorically means that the extraction as well as cosumption of naturally available resources need to decouple from the economic growth. For perspective; non-renewable resources with negative footprints (fossil fuels, difficult to recycle materials) can lead to big "competitive disadvantage" in the future. Therefore, the "circular" approach can only keep the resources in use for the economy. Let us be clear and loud: sustainability is only possible by adoption of "Circular economy". Technologically, it is nothing but turning waste into wealth and bringing back to the manufacturing. Therefore, this will allow us to close the loop for supply chain in ideal case with zero waste. We can solve both the problem of supply chain of various raw materials and environmental impact of waste disposal.

8.3 STRATEGIES FOR DEVELOPMENT OF SUSTAINABLE WASTE MANAGEMENT TECHNOLOGIES

As mentioned in the section 8.2, adoption of circular economy with zero wastage is the best strategy for sustainable waste management. Waste needs to be fully converted to wealth economically and technologically. The waste provides possibly the biggest economic opportunity in our time. Applying technological solutions to turn waste into wealth can not only solve the problem of "rubbish" but also make financial gain, enabling sustainable growth without stressing the constrained resources. The following strategies can be adopted to take advantage of the circular economy.

(a) Adoption of circular supply chain
This is the heart of the circular economy. This will require us to introduce completely renewable, recyclable, reusable, and biodegradable materials, which can further be utilized again and again. For example, the technology of mimicking the way plants use photosynthesis to prepare food can be adopted to replace all fossils-based raw materials.

(b) Reuse, recover, and recycle
This will allow everything which is considered waste to be reused. The necessary secondary materials can be recovered or recycled to the production stream. The energy components can be reclaimed.

(c) Extension of product
We normally discard any products which we find no longer useful. The products may be damaged, out of fashion, or no longer required. However, they can economically be useful if we maintain and improve them by repair, remanufacture or remarket. This provides the avenues for keeping the products relevant and useful for many cycles.

(d) Sharing a common platform
This is like the Google store, where all our necessary Apps are available. The sharing platform can enable us to forge new relationship and opportunities for all-business houses, consumers, entrepreneurs, and waste recovers, bringing all the parties involved manufacture and use of a product for sharing.

(e) Product as service
Instead of the "buy, use and throw" model, we can use the product as service. Any consumer can lease or pay for a product as a service model. After use, the product goes back to the business houses for repair, remodeling, refurbishment, and reuse again. Major airline companies use this concept for sustainability.

8.4 TECHNOLOGICAL WASTE MANAGEMENT: ACCOUNTABILITY, RISK AVAILABILITY, AND RESILIENCE

The multinational companies have been using the linear model of economics for the last three centuries leading to the generation of massive amount of waste. Each individual in his/her capacity generates waste and it is his/her responsibility to take care of safe disposal, treatment and further reuse. However, in reality, we treat waste differently from the useful objects. Hence, heaves of waste have been mounting, without much treatment to reuse or recycle. We all need to be aware of the consequences of unsafe disposal of waste. This is true for all types of waste across all domains; household, industry, hospitals, construction houses even schools, colleges etc. In fact, it is the responsibility of each individual to make the waste organized efficiently. Responsibility begins at home: segregation of waste (solid vs. liquid, bio-vs. plastic, metal vs. glass etc.) followed by safe transfer to waste management team in their locality. This needs to be followed by every individual in villages or town or cities. In a nutshell, we all need to take responsibility for waste generated. We must not forget that waste is not just the responsibility of someone who manages it. United Nations guidelines describe a waste producer as someone:

- who made the substance became waste
- who decided that a substance was unwanted and hence, waste.

However, waste management involves multi-faceted challenges. It ranges from generation of waste, segregation, collection, transport, treatment, reuse, recycle and disposal. The solution to this problem must involve all stakeholders at diverse levels – household, businessmen, manufactures, government, policy makers etc. The sole responsibility cannot only lie with local and national governments. Private players; companies need to exercise the "cleanup" exercise in the form of corporate social responsibility (CSR). As the problem of waste management is complex, all stockholders must realize an urgent need to collaborate and form a partnership and task force. An example of this partnership is related to PET bottles. Worldwide partnerships and task forces are needed for effective disposal, reuse, and recycling of PET bottles, which is considered as an environmental menace. Another model known as the take-back-scheme is based on involving all stakeholders in the entire value chain; factory-customer-collector-recycler. This way, it will be effective. Such a strategy has already been implemented in countries like Germany, where PET bottles are being "taken back" and customers are paid back the price of the bottle. This way, one can ensure collection of the PET for safe disposal, recycling, or even reuse.

The waste management also involves risks. In order to make this reliable and effective, we need to work on the risks and sort them out effectively. The risks can be global or local. Here we shall discuss global risks. Although we vouch for adaptation of a circular economy with zero wastage, it seems to be a distant dream. Linear economy, which thrives on available resources, is difficult to transform to circular in a quick time frame. Therefore, the risk of resource scarcity and environmental degradation as well as climate change is immediate; needing urgent attention. With a large population and rising demand for resources, we have of risk of procrastination of measures, leading to irreparable and permanent damage to the Earth. The second risk at the global level is related to value chain efficiency. It must be mentioned that the base of the circular economy cannot be based on "one size fits all" approach. This involves the risk of falling out of a circular economy. We need to work in unison to make the progress of a circular economy, which is acceptable to all.

8.5 FUTURE DIRECTION

The future of waste management is definitely the circular economy- which will provide a holistic solution for sustainability. Adoption of a circular economy will provide many advantages. Many multinational companies have been making serious efforts to adopt the circular economy. In fact, there is considerable urgency to move towards circular. However, the path is strewn with many challenges. These challenges can be surmounted by adopting efficient manufacturing processes, value engineering, and

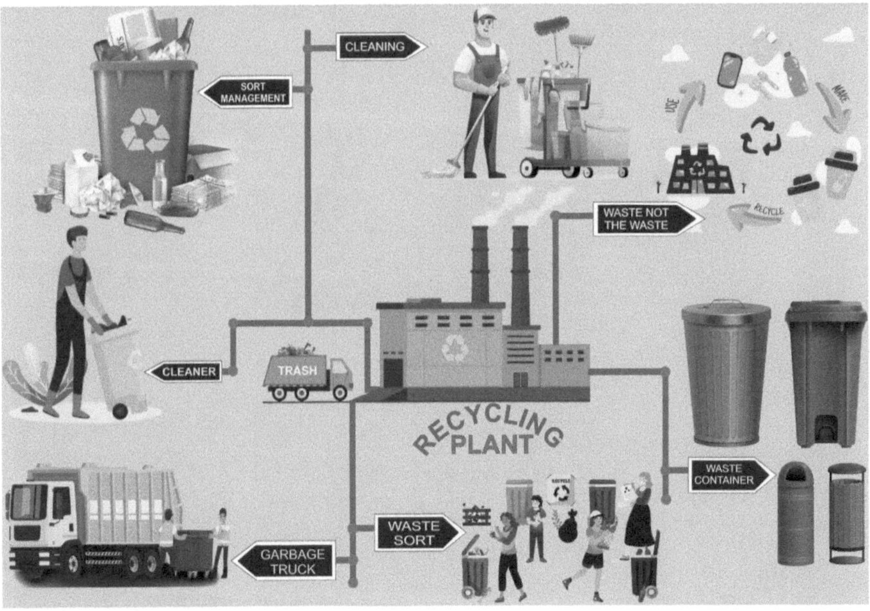

FIGURE 8.2 Application of AI in waste management. (From Sanksshep Mahendra in https://aiplusinfo.com/blog/artificial-intelligence-in-waste-management/)

usage of digital technology and AI. Each of the circular economic models offers a unique means to add value to a product or a process. However, we need to find the best path. This will depend on understanding how to use more business models to protect the value. Finding new ways to add value may not come easy. Rapid development of digital technology in the last few decades is considered a major enabler for a circular economy. This will allow us to adopt mobile and advanced technologies for resource flow organization, supply chain, effective engagement of consumers, offering products and service. AI (Artificial Intelligence) can be put into use in almost every step of a circular economy. The explosive growth of AI in recent years has made us believe that it has the potential to change almost everything we perceive. The cartoon shown in Figure 8.2 captures the application of AI in waste management. This offers game-changing options, providing all possible solutions to the waste mangement and reducing risk to bare minimum.

REFERENCES

[1] UNEP, "Responsible Resource Management for a Sustainable World: Findings from the International Resource Panel," 2012, www.unep.org/resource-panel/Portals/24102/PDFs/SYNOPSIS%20Final%20compressed.pdf

[2] Statistics Brain, "World Poverty Statistics," November, 2014, www.statisticsbrain.com/world-poverty-statistics/

Index

Note: Page numbers in *italics* indicate figures and in **bold** indicate tables on the corresponding pages.

R

S

T

U

V

W

Y

Z

FSC
www.fsc.org
MIX
Papier | Fördert gute Waldnutzung
FSC® C083411

Kolibri 360 GmbH
Ferdinand-Jühlke-Straße 7
99095 Erfurt, Deutschland
produktsicherheit@kolibri360.de